INCLINE ALGEBRA AND APPLICATIONS

INCLINE ALGEBRA AND APPLICATIONS

Z.-Q. CAO
Institute of Automation,
Academia Sinica, China

K.H. KIM, B.S., M.S., M.Phil., Ph.D.
Kiro Research Group,
Montgomery, Alabama.

F.W. ROUSH, A.B., Ph.D.
Kiro Research Group,
Montgomery, Alabama.

ELLIS HORWOOD LIMITED
Publishers · Chichester

Halsted Press: a division of
JOHN WILEY & SONS
New York · Brisbane · Chichester · Toronto

First published in 1984 by

ELLIS HORWOOD LIMITED
Market Cross House, Cooper Street, Chichester, West Sussex,
PO19 1EB, England

*The publisher's colophon is reproduced from James Gillison's drawing of
the ancient Market Cross, Chichester.*

Distributors:
Australia, New Zealand, South-east Asia:
Jacaranda-Wiley Ltd., Jacaranda Press,
JOHN WILEY & SONS INC.,
G.P.O. Box 859, Brisbane, Queensland 4001, Australia.

Canada:
JOHN WILEY & SONS CANADA LIMITED
22 Worcester Road, Rexdale, Ontario, Canada.

Europe, Africa:
JOHN WILEY & SONS LIMITED
Baffins Lane, Chichester, West Sussex, England.

North and South America and the rest of the world:
Halsted Press: a division of
JOHN WILEY & SONS
605 Third Avenue, New York, N.Y. 10158, U.S.A.

©1984 Z-Q. Cao, K.H. Kim and F.W. Roush/Ellis Horwood Ltd.

British Library Cataloguing in Publication Data

Cao, Z-Q.
Incline algebra and applications. –
(Ellis Horwood series in mathematics and its applications)
1. Algebra
I. Title   II. Kim, Ki Hang
III. Roush, F.W.
512   QA155

Ellis Horwood Edition ISBN 0-85312-765-4
Halsted Press ISBN 0-470-20116-9
Ellis Horwood Student Edition ISBN 0-85312-818-9

Printed in Great Britain by Unwin Brothers of Woking.

Table of Contents

Preface

This is the first attempt to collect published and unpublished results on incline algebra, which encompasses Boolean algebra and fuzzy algebra.

In order to read this monograph, the reader should have had the usual first year graduate courses in mathematics plus an elementary knowledge of combinatorial theory, graph theory, lattice theory, matrix theory (including the Boolean and fuzzy cases), semi-group theory, and semiring theory.

In accordance with Occidental practice, the surname is placed last in all Oriental names in this book.

The authors gave lectures on portions of this book to: Institute of System Sciences, Academia Sinica, Graduate School of Academia Sinica, Graduate School of China University of Science and Technology, Beijing Normal University, Northwestern Normal University – Changchun, Yanbian University – Yanji, Institute of Automation, Academia Sinica, Korean Academy of Sciences – Pyongyang, and Alabama State University – Montgomery.

The authors gratefully acknowledge the following persons for encouragement, criticism, and suggestions: Jingqing Han, Hiroshi Hashimoto, Shanyu He, Kuochih Hsu, Chaochih Kwan, Joseph Neggers, Boris M. Schein, Maurice Tchuente, Peizhuang Wang, Elliot S. Wolk, Jici Yan, and Jinwen Zhang.

Lastly, the authors are indebted to Mrs. Kay Roush for very diligent and accurate proofreading of the manuscript.

August 15, 1984

Z.–Q. CAO Beijing, CHINA
K. H. KIM Mundok, KOREA
F. W. ROUSH Montgomery, Alabama

Introduction

In the 1800's mathematicians discovered that propositional logic could be represented by a new structure called *Boolean algebra* in which $0 + 0 = 0$, $1 + 0 = 0 + 1 = 1$ but $1 + 1 = 1$, where 1 is true and 0 is false. This representation can be extended from propositions like p and (q or r) to the more intricate logic of binary relations by taking matrices over the Boolean algebra. That is, let a relation "x < y" be represented on a set $\{x_1, \ldots, x_n\}$ by a matrix A where $a_{ij} = 1$ if $x_i < x_j$ and $a_{ij} = 0$ if $x_i \not< x_j$. With natural definitions transitivity becomes the matrix inequality $A^2 \leq A$.

In the 1960's this was extended to a kind of multivalued logic called *fuzzy sets*. Here we take as elements all real numbers between 0 and 1 and add by max $\{a, b\}$, multiply by min $\{a, b\}$. This allows us to represent relationships in which there is a notion of degree to which the relationship holds and the degree of a composition is the minimum of the degrees of its

components. For example, "x_1 is an associate of x_2"
with degree 0.6 can be stated as a_{12} = 0.6.

Boolean algebra and the theory of fuzzy sets are
two examples of a general structure called *incline*.
In a general incline, the degree of relationship in a
composition may be less than either factor. One example is obtained when we take as elements the interval
[0, 1] again and operations max {a, b}, and xy. This
combines arithmetic and ordered structures.

In an incline the additive operation is max {x, y}
taken within an order structure and the multiplicative
operation can be any associative distributive operation
such that xy is less than or equal to either factor.
This gives the reason for the name: under repeated
operations, quantities tend to decrease.

Inclines can also be used to represent automata and
other mathematical systems, in optimization theory, to
study inequalties for nonnegative matrices and matrices
of polynomials. The two-sided ideals of any ring form
an incline.

Incline theory is based on *semiring theory* and
lattice theory. A *semiring* is a set provided with two
binary operations such that for all elements a, b, c;
a + b = b + a, a + (b + c) = (a + b) + c, a(bc) = (ab)c
a(b + c) = ab + ac, (b + c)a = ba + ca. A semiring
homomorphism is a mapping f: $R_1 \rightarrow R_2$ such that for all
a, b; f(a + b) = f(a) + f(b), f(ab) = f(a)f(b). Matrices over any semiring again are a semiring. The nonnegative integers is a semiring. Homomorphisms of semirings are studied by means of equivalence relations
preserved by multiplication on either side, called
congruences.

A *partially ordered set* (poset) is a set on which
a relation "<" is defined such that a $\not<$ a and if a < b,

b < c then a < c. *Semilattices* are posets in which any two elements have a least upper bound. Then least upper bound gives an associative, commutative operation with a + a = a. Every finite semilattice can be represented as a family of sets with the least upper bound being union of the sets.

A *lattice* is a semilattice in which any two elements also have a greatest lower bound. This gives a second operation. If the operations are distributive over one another the lattice is called *distributive*. Every distributive lattice is an incline.

The reader can probably learn most of the semiring theory and lattice theory he needs as he goes along. For semiring theory, see Clifford and Preston (1961) on congruences, generators, relations, which are the same as for semigroups. For lattice theory, see the elementary material in Birkhoff (1967).

Special cases of incline theory have appeared in Cuninghame-Green's *Minimax Algebra* (1979) and in U. Zimmerman (1981).

Chapter 1 of the present book is on the theory of commutative inclines in themselves. We show that in inclines (later generalized to the noncommutative case) every infinite chain has a subsequence with $a_i \geq a_j$ if $j > i$. This means finitely generated inclines have a lattice structure.

Conversely, finite lattices have a nonzero incline structure (with new multiplication) if they admit a nonzero distributive operation. We also study linear representability, later also real representability (is an incline a subincline of a product of linearly ordered inclines ?) and characterize incline structures over finite Boolean algebras and lattices of subspaces of a vector space.

Chapter 2 introduces vectors and matrices. We show that invertible matrices must be generalized permutation matrices. We prove existence and uniqueness theorems for bases of a subspace of a vector space.

Chapter 3 is mainly on idempotents (matrices satisfying $A^2 = A$) and generalized inverses. We can reduce idempotents over a finite incline so that main diagonal elements are idempotent and nonzero rows form a basis. This implies equality of row and column rank when defined. There exists a weak triangular form. We give algorithms for finding generalized inverses B of A, where ABA = A. If a matrix has a Moore-Penrose inverse, then its transpose is a Moore-Penrose inverse.

An *antiincline* differs from an incline in that $xy \geq x$, $xy \geq y$. We extend various results to noncommutative inclines and antiinclines.

In Chapter 4 on topological inclines we show that every incline structure on [0, 1] in which $1^2 = 1$ can be built up from the fuzzy algebra and two other structures. A compact metric topology on an incline is determined by its ordering. We discuss n-dimensional topological inclines, characterize 1-parameter semigroups of matrices over an incline on [0, 1], and study infinite dimensional vector spaces over an incline.

In Chapter 5 we study the sequence of powers of a matrix. We give a formula for $\lim_{n \to \infty} A^n$ if it exists. Sets of matrices forming a group go isomorphically to groups of Boolean matrices in many cases. The group-theoretic complexity of the set of n × n matrices over an incline is n - 1 if the incline is a subincline of a direct product of inclines whose only idempotents are 0, 1.

Introduction

The last chapter gives a survey of applications: models of the brain, interactions of political systems, automata, dynamic programming, social choice, geodesics in graphs, clustering, matrix theory, assignment problems.

A list of open problems is given at the end of each chapter.

Chapter 1

Incline Algebra

An *incline* is a structure which has an associative, commutative addition, and distributive multiplication such that x + x = x, x + xy = x for all x, y. It has a semiring structure and partial order structure. In this chapter we consider the standard concepts: sub-algebra, quotient algebra, ideal, congruence, direct product, generators, relations for inclines. These apply in a way similar to their use for groups and semigroups.

Every incline is a semilattice. Our main concerns in this chapter are (1) when is an incline a lattice ? a distributive lattice ? (2) when does a lattice have a nontrivial incline structure ? (3) when can an in-cline be represented by simpler inclines ?

1.1. BASIC PROPERTIES OF INCLINES

An *incline* is an algebraic structure with two opera-tions, addition and multiplication. It generalizes distributive lattices in that multiplication need not be idempotent.

We have *Boolean algebras* ⊂ *fuzzy algebras* ⊂ *distributive lattices* ⊂ *inclines* ⊂ *semirings*. Therefore, it may have applications to new models in various sciences. A related structure called *blog* was defined and studied by Cuningham-Green (1979). U. Zimmerman (1981) studied *idempotent semirings*. We summarize their work at the end of this chapter.

DEFINITION 1.1.1.   An *incline* is a set S and two binary operations +, · on S such that for all x, y, z in S; (1) *(Associative)*: x + (y + z) = (x + y) + z

$$x(yz) = (xy)z$$

(2) *(Commutative)*: x + y = y + x,   xy = yx

(3) *(Distributive)*: x(y + z) = xy + xz

(4) *(Idempotent)*:   x + x = x

(5) *(Incline)*:   x + xy = x

Note that these operations tend to make quantities "*slide downhill*." Therefore, we decided to call it *incline* and let **7** denote an arbitrary incline where **7** is the first letter of the *Korean alphabet* and is pronounced as "*Gee-Uck*" and it resembles an incline.

EXAMPLE 1.1.1.   The Boolean algebra {0, 1} is an incline under Boolean operations.

EXAMPLE 1.1.2.   The fuzzy algebra [0, 1] is an incline under the operations maximum and minimum.

PROPOSITION 1.1.1.   (Cao). *Every distributive lattice is an incline. An incline is a distributive lattice (as semiring) if and only if $x^2 = x$ for all x.*
    Proof.   The first statement is immediate. If y ≤ z then y + z = z, so x(y + z) = xz so xy ≤ xz. If any

distributive lattice $x \wedge x = x$. Conversely we show that if $x^2 = x$ holds then $xy = x \wedge y$ always. We have $xy \leq x$ by $x + xy = x$. By commutativity $xy \leq y$. Suppose $u \leq x$ and $u \leq y$. Then $u = u^2 \leq xy$. This proves $xy = x \wedge y$. $\square$

EXAMPLE 1.1.3.   The set $[0, 1]$ under the usual multiplication and partial order is an incline but does not satisfy $x^2 = x$.

It follows immediately from the definition that an incline is a semilattice under addition and a semigroup under multiplication. Any unspecified ordering in an incline is assumed to be the semilattice ordering.

PROPOSITION 1.1.2.   *Every incline is a semilattice ordered commutative semigroup in which $xy \leq x$, under the same operations.*

Proof.   Let $x \leq y$ in an incline. Then $zx + zy = z(x + y) = zy$. Therefore $zx \leq zy$. This proves we have a semilattice ordered commutative semigroup.

The converse is false since a nondistributive lattice is a semilattice ordered commutative semigroup in which $xy \leq x$ but is not an incline. $\square$

PROPOSITION 1.1.3.   *In the multiplicative semigroup of an incline all Green's J-classes are points.*

Proof.   This is true in any ordered semigroup in which $xy \leq x, y$. $\square$

The concepts of generators, relations, subinclines, homomorphisms, congruences, ideals, are the same as for semirings.

DEFINITION 1.1.2.   A *subincline* of an incline 7 is a subset closed under addition and multiplication.

EXAMPLE 1.1.4.   Any subset of [0, 1] under the real number product is a subincline if and only if it is closed under multiplication.

DEFINITION 1.1.3.   An *ideal* in an incline 7 is a subincline 7' ⊂ 7 such that if x ε 7' and y < x then y ε 7'.

EXAMPLE 1.1.5.   For any x, the set {a: a ≤ x} is an ideal.

DEFINITION 1.1.4.   The subincline *generated* S is the set of all elements which can be obtained by adding and multiplying finite sequences of elements of S.

DEFINITION 1.1.5.   A subincline generated by a single element is *cyclic*.

EXAMPLE 1.1.6.   The negative integers form a cyclic incline generated by -1.

DEFINITION 1.1.6.   A *congruence* on an incline 7 is an equivalence relation ~ on 7 such that if a ~ h then ax ~ hx and $a + x$ ~ $h + x$ for all a, h.

EXAMPLE 1.1.7.   All congruences on the negative integers have the form x ≡ y if x = y or if x, y are both less than k for some k.   (Compare Example 1.1.6.)

DEFINITION 1.1.7.   A *homomorphism* of inclines is a mapping $f: 7 \to S$ such that $f(x + y) = f(x) + f(y)$ and $f(xy) = f(x)f(y)$.

EXAMPLE 1.1.8.   There exists a homomorphism from the negative integers under addition to $[0, 1]$ sending n to 2^n.

For brevity, let Z^- denote the set of all negative integers.

DEFINITION 1.1.8.   The incline $Z_m^- = \{-1, -2, \ldots, -m\}$ in which the operation is sup $\{x + y, -m\}$.

PROPOSITION 1.1.4.   *Every cyclic incline is isomorphic to* Z^- *or* Z_m^-.

Proof.   The elements of a cyclic incline are all powers of the generator x since $x^n + x^m = x^n$ if $n \leq m$. Thus they are $x, x^2, \ldots$. We have $x \geq x^2 \geq \ldots$. If equality never holds we have Z^-. If $x^m = x^{m+1}$ then $x^m x = x^{m+1} x$ so $x^m = x^{m+2}$ by transitivity. So $x^n = x^m$ for $n > m$ and we have Z_m^- for m minimal.   □

DEFINITION 1.1.9.   The *quotient incline* given by a congruence is the incline formed by the equivalence classes under the operations $\overline{x} + \overline{y} = \overline{x + y}$, $\overline{x}\,\overline{y} = \overline{xy}$, which are well defined.

EXAMPLE 1.1.10.   The incline Z_m^- is a quotient of Z^-.

PROPOSITION 1.1.5.   *If* $f: 7 \to S$ *is an incline homomorphism and* $x \sim y$ *is the equivalence relation* $f(x) = f(y)$ *then* $x \sim y$ *is a congruence and the image* *of f is isomorphic to the quotient incline.*

   Proof.   If f(x) = f(y) then f(x + z) = f(y + z) and
f(xz) = f(yz).   Since x ∽ y if and only if f(x) = f(y)
image elements f(x) are in 1-1 correspondence with
elements of the quotient incline.   □

   EXAMPLE 1.1.11.   Let S be any semilattice with 0.
Then S is an incline under the operation xy = 0.

   Semilattices with 0 need not be lattices.

   EXAMPLE 1.1.12.   Let S be the semilattice consisting of [0, 1) together with x, y, x + y such that the
partial order is given by x > t, y > t for t ∈ [0, 1)
and x + y > x, x + y > y, with the usual order on
[0, 1).   Then x, y have no infimum.

   F. Ramsey (1930) proved the basic theorem of infinite Ramsey theory, in a set-theoretic context.   His
theorem states that *if S is an infinite set and the se*
⊖(S) of unordered pairs from S is colored with a
finite set of colors, there exists an infinite subset
T such that ⊖(T) has only one color. We can restate
this theorem in Boolean or nonnegative matrix form.

   THEOREM 1.1.6.   (Ramsey).   *Let M be an infinite sym*
metric matrix over the Boolean algebra {0, 1} or over
a finite subset of **Z**. *Here* **Z** *denotes the set of all*
integers.   Then there exists a principal infinite
submatrix A all of whose off-main diagonal entries
are equal.

   Note that if M is reflexive over the Boolean algebr
{0, 1}, then either A = I or A = J where I denotes

an identity matrix and J denotes a universal matrix.

EXAMPLE 1.1.13.   Let M be the same as in Theorem
1.1.6. Let $m_{ij} = 1$ if and only if $i|j$ or $j|i$. The
principal submatrix such that i, j are powers of 2
equals J.

This theorem implies the following result for
inclines.

THEOREM 1.1.7.   *For a finitely generated incline 7*
for any sequence $w_i \in 7$, *i* $\in$ *Z, there exists i < j*
with $w_i \geq w_j$.

Proof.   It suffices to prove this in the case when
7 is the incline of sums of monomials in n generators,
since there is an epimorphism from this incline to any
incline with n generators.

Suppose the theorem is false for a sequence w_i.
Then for i < j since $w_i \not\geq w_j$, w_j has a monomial m_{ij}
not less than or equal to any monomial of w_i. Let m_{10}
be any monomial of w_i. Then we can replace w_i by
$\sum_j m_{ij}$ and w_1 by m_{1j} and still $w_i \not\geq w_j$ for i < j. We
have m_{ij} has a lower power of some variable than the
highest power occurring in w_i. Form a matrix where
the (i, j)-entry is k where variable k is such a
variable. Apply Ramsey's theorem. We get a subsequence for which the m_{ij} has a lesser power of a fixed
variable k, than does w_i. Let v_i be $\sum_j m_{ij}$ restricted
to this subsequence. Then variable k cannot occur to
a higher power in v_i than its highest power in v_{i-1},
v_{i-2}, $\cdots$, v_1. So the powers of variable k occurring
are bounded. Now apply Ramsey's theorem again, with
a matrix whose (i, j)-entry is k if m_{ij} has that

variable to a power k. Then $\sum_j m_{ij}$ over this
subsequence gives a new sequence, with that variable
occurring to a constant power. Now we have reduced
the problem to a problem in n - 1 generators. For
n = 1, the result is immediate. So by induction the
proof is complete.   □

THEOREM 1.1.8.   *An incline is a lattice under any*
of the following conditions: (i) intervals are finite;
(ii) it is finitely generated over +; (iii) + is con-
tinuous in a compact topology; (iv) each interval
lies in a subincline finitely generated over + and ·;
and (v) + is continuous and intervals are compact.
 Proof.   Under (i), (iii) the quantity sup $\{z: z \le x$
and $z \le y\}$ defines $x \wedge y$. Since xy is in the set, it
is nonempty. For (iv), (v) the result will follow
from (ii), (iii) by considering the interval [xy, x +
y]. If $u \le x$ and $u \le y$ then u + xy is a quantity in
this interval with the same property, and u + xy $\ge$ u.
So a greatest lower bound in [xy, x + y] will be a
greatest lower bound in the whole lattice.
 It remains to prove (ii). We will consider the set
$\{z: z \le x, z \le y, z \ge xy\}$. By the Hausdorff maximal
principle to show that the supremum of $z \ge xy$ such
that $z \le x$, $z \le y$ exists it will suffice to show
that every increasing chain in this set has an upper
bound. The result now follows from Theorem 1.1.7.   □

PROPOSITION 1.1.9.   *A finitely generated incline*
has no infinite ascending chain.
 Proof.   This follows directly from Theorem 1.1.7. □

EXAMPLE 1.1.14.   The incline [0, 1] has the infinite ascending chain $\{1 - \frac{1}{n}\}$.

PROPOSITION 1.1.10.   In a finitely generated incline there exists no infinite ascending chain of ideals.

Proof.   In each ideal we can pick an element not less than or equal to any element of a previous ideal.☐

Every element of an incline can be expressed as a polynomial in the generators (without coefficients) in which no monomial divides another since if $M_1 | M_2$ then $M_1 \geq M_2$ so $M_1 + M_2 = M_1$. A relation is an equation $\Sigma M_i = \Sigma M_j$ of such polynomials. The congruence generated by a set of relations is the smallest congruence in which those relations hold.

PROPOSITION 1.1.11.   The congruence generated by a set of relations $w_i \sim z_i$ is the transitive closure of the relation $\mathfrak{R}$ such that $(az_i + b) \ \mathfrak{R} \ (aw_i + b)$ and $(aw_i + b) \ \mathfrak{R} \ (az_i + b)$ where symbolically, we allow $z + b \sim w + b$ (as if $a = 1$).

Proof.   This relation must be contained in the congruence. Conversely it will suffice to show it is a congruence. It is reflexive (take b = a), symmetric, and transitive. If we add any element or multiply any element to $(az_i + b) \ \mathfrak{R} \ (aw_i + b)$ we obtain a relation of the same form. Thus this is a congruence.   ☐

DEFINITION 1.1.10.   The incline with generating set T and defining relations $u_i = w_i$ is constructed as follows. Take the set of all monomials

$$\prod_i x_i^{n_i}$$

for x_i in T. This is a commutative multiplicative

semigroup. Take the family of all finite formal sums of monomials $m_1 + m_2 + \ldots + m_n$ subject to the semigroup relation $m_1 + m_2 = m_1$, if $m_1 | m_2$. This gives an incline. Now take the congruence generated by the relations $u_i = w_i$. Then the equivalence classes form the desired incline.

EXAMPLE 1.1.15. For two generators x, y and no relations, this incline is the semiring of all polynomials $\Sigma\ x^r y^s$ subject to $m_1 + m_2 = m_1$ if $m_1 | m_2$.

PROPOSITION 1.1.12. *For any incline 7 and mapping $f : S \to 7$ such that the elements $f(s)$ satisfy the relations $u_i = w_i$ there exists a unique incline homomorphism $\overline{f}: T \to 7$ such that $\overline{f} = f$ on S, where T is the incline with defining relations $u_i = w_i$.*

Proof. Define f_1 by $f_1(m_1 + m_2 + \ldots + m_n) = \overline{f}(m_1) + \overline{f}(m_2) + \ldots + \overline{f}(m_n)$ and

$$f_1(\Pi\ x_i^{n_i}) = \Pi\ f(x_i)^{n_i}$$

Then f_1 induces the required homomorphism $\overline{f}$ on equivalence classes. $\square$

1.2 INCLINE STRUCTURES ON LATTICES

Every semilattice with 0 has an incline structure given by $xy = 0$.

PROPOSITION 1.2.1. *On any finite lattice there exists a unique maximal binary operation xy such that $xy \le x$ and $xy \le y$ and $x(y + z) = xy + xz$ and $(y + z)x = yx + zx$. This operation is commutative.*

Proof. If x o y and x * y are operations of this kind, so are x o y + x * y and x o y + y o x. Hence we can take the sum of all such operations of this kind which will be the unique maximal one. If it were not commutative xy + yx would be greater. ☐

We do not know whether this operation is always associative. It does give an upper bound on incline structures. Moreover, we will later show that if it is nonzero there exists a nonzero incline structure.

There exists an effective way to compute this maximal distributive operation on any finite lattice. Start with $xy = x \wedge y$. Then whenever $x(y + z) > xy + xz$ redefine $x(y + z) = xy + zx$, $uv = \inf \{uv, xy + xz\}$ for $u \le x$, $v \le y + z$. This process must terminate in finitely many steps since it always decreases the operation. The product must be greater than or equal to the maximal product at each stage. But since the result will be distributive it must equal the maximal distributive product. The product remains monotone at each step.

EXAMPLE 1.2.1.

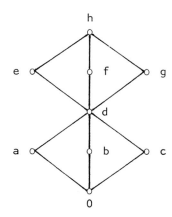

We have $h^2 = (e + f)(e + g) \le e + fg = e$. So re-define $h^2 = e$, $f \cdot f = g \cdot g = d$, $h \cdot f = h \cdot g = d$. Also $h^2 = (f + g)(f + g) = d$. Redefine $x \cdot y = d$ for $x, y > d$. By a similar argument redefine $x \cdot y = 0$ for $y \le d$. This gives a distributive operation, which is maximal. It is associative and is therefore an incline structure.

PROPOSITION 1.2.2.   *In any incline, the set $\{x: x = x^2\}$ is a subincline which in itself is a distributiv lattice.*

Proof.   It suffices to prove that it is closed under addition and multiplication, by Proposition 1.2.1. If $x^2 = x$, $y^2 = y$ then $(xy)^2 = x^2y^2 = xy$. And $xy = x^2y^2 \leq x^2$. Therefore $(x + y)^2 = x^2 + xy + y^2 = x^2 + y^2 = x + y$.   □

EXAMPLE 1.2.2.   For the fuzzy structure on $[0, 1]$ every element satisfies $x^2 = x$. For the structure xy, only 0, 1 do.

PROPOSITION 1.2.3.   *Let π be a function from a lattice to a sublattice which is distributive and sucl that $\pi(x + y) = \pi(x) + \pi(y)$, $\pi(xy) = \pi(x)\pi(y)$, $\pi\pi(x) = \pi(x)$ and $\pi(x) \leq x$. Then $xy = \pi(x)\pi(y)$ gives an inc-line structure in which $xy = (xy)^2$. Conversely every incline structure in which $xy = (xy)^2$ has this form.*

Proof.   The first statement follows from the defi-nition. Suppose that $xy = x^2y^2$. Then by the result above the set of elements of the form xy is a distri-butive sublattice. Let $\pi(x) = x^2$. We have $\pi(x) \leq x$, and $\pi(xy) = x^2y^2 = xy$, $\pi(x + y) = x^2 + xy + y^2 = x^2 + x^2y^2 + y^2 = \pi(x) + \pi(y)$, $\pi(\pi(x)) = x^4 = x^2x^2 = xx = \pi(x)$.   □

DEFINITION 1.2.1.   If $xy = x^2y^2$ in an incline, we call it a π-*incline*.

EXAMPLE 1.2.3.   Boolean vectors have a π-incline structure in which $(a, b)(c, d) = (0, bd)$.

THEOREM 1.2.4.   *The semilattice of n-dimensional Boolean vectors V_n has exactly 2^n incline structures, which are all π-inclines.*

Proof.   There exist 2^n π-inclines defined by $\pi(x)$ = xv for any Boolean vector v. Let xy be any incline structure. Let e_1, e_2, ... , e_n be the basis for V_n. Then $e_i e_j \leq e_i \wedge e_j = 0$, for $i \neq j$ and so $e_i e_j = 0$ for $i \neq j$. And $e_i^2 \leq e_i$ so $e_i^2 = e_i$ or $e_i^2 = 0$. Define v by $v_i = 1$ if and only if $e_i^2 = e_i$. Let $\pi(x) = xv$. This gives a π-incline. $\square$

DEFINITION 1.2.2.   An incline structure is *nil* if $x^3 = 0$ for any x. Since $(x + y + z)^3 = 0$, this implies $xyz = 0$ for all x, y, z since $xyz \leq (x + y + z)^2$.

THEOREM 1.2.5.   *Any finite semilattice admits a nontrivial incline structure if and only if it admits either a nontrivial π-incline structure or a nontrivial nil incline structure.*

Proof.   Suppose we have a nontrivial incline structure. Suppose that $a^2 = a$ for some $a \neq 0$. Choose n large enough that $x^n = x^{n+1}$ for all x in the semilattice. Then the product $x^n y^n$ satisfies $x^n y^n \leq xy \leq x$, $(x^n y^n)^n z^n = x^n y^n z^n = x^n (y^n z^n)^n$, $x^n y^n = y^n x^n$, $x^n (y + z)^n = x^n (y + z)^{2n} \leq x^n y^n + x^n z^n$ by expansion of $(y + z)^{2n}$. But $x^n (y + z)^n \geq x^n y^n + x^n z^n$ is also true. This gives a nontrivial π-incline structure.

Suppose $a^2 < a$ for all $a \neq 0$. Choose the least n such that $x^n = 0$ for all x. Then $x_1 x_2 ... x_n = 0$ also since $x_1 x_2 ... x_n \leq (x_1 + x_2 + ... + x_n)^n$. We have $n \geq 3$. Let m be the maximal element of the semilattice.

Define $x * y = xym^{n-3}$. Then this is a nil incline. Associativity holds by $(x * y) * z = 0$, since there are at least n factors. □

THEOREM 1.2.6. *A finite lattice has a nontrivial π-incline structure if and only if there exists a nonzero element v such that if $y + z \geq v$ then $y \geq v$ or $z \geq v$.*

Proof. Suppose v exists. Define a product by $xy = v$ if $x \geq v$ and $y \geq v$ and $xy = 0$ otherwise. Then $xy \leq x$, and $xy = yx$. Suppose $x(y + z) > xy + xz$. Then $x(y + z) = v$, $xy = 0$, $xz = 0$. Then $x \geq v$, $y + z \geq v$. So $y \geq v$ or $z \geq v$. So $xy = v$ or $xz = v$. This is false. So distributivity holds. Suppose $x(yz) \neq (xy)z$. By symmetry, let $x(yz) = v$, $(xy)z = 0$. Then $x \geq v$, $yz \geq v$ so $y \geq v$, $z \geq v$. So $(xy)z = v$. This is false. So this gives a π-incline structure.

Conversely, suppose we have a π-incline structure. Let v be a basis element for the image of π, which is distributive. Suppose $x + y \geq v$. Then $\pi(x) + \pi(y) \geq v$. So $v = v\pi(x) + v\pi(y)$. So $v = v\pi(x)$ or $v\pi(y)$. So $x \geq v$ or $y \geq v$. □

EXAMPLE 1.2.4. Suppose a lattice has unique minimal nonzero element. Then it has a π-incline structure.

THEOREM 1.2.7. *A finite lattice admits a nonzero incline structure if and only if there exists a nonzero distributive product such that $xy \leq x$ and $xy \leq y$.*

Proof. If an incline structure exists, this gives a nonzero distributive product. Suppose the maximal distributive product xy satisfying $xy \leq x$ and $xy \leq y$ is nonzero. If there exists v such that if $x + y \geq v$

then $x \geq v$ or $y \geq v$ then we have a π-incline. Suppose this is false. We next show that $x^2 < x$ for all $x > 0$ in the lattice. First suppose x is a lattice basis element. Write $x \leq y + z$ where $y \ngeq x$, $x \nleq z$. Then $x^2 \leq x \wedge (y + z) = x \wedge y + x \wedge z$. Both $x \wedge y < x$ and $x \wedge z < x$. Since x is a basis element it cannot be a sum of strictly smaller vectors. So $x \wedge y + x \wedge z \neq x$, so $x^2 \leq x \wedge y + x \wedge z < x$. Next suppose x is not a basis element. Write it as a sum $x_1 + x_2 + \ldots + x_n$ of basis elements which is minimal in the sense that if $y_i \leq x_i$ and $\Sigma\, y_i = x$ then $y_i = x_i$ for each i. Such a minimal expansion exists since if any expression is not minimal we can replace it by a smaller one. Then

$$x^2 = (x_1 + x_2 + \ldots + x_n)^2 \leq \Sigma\, x_i^2 + \sum_{i<j} x_i \wedge x_j \leq$$

$x_1^2 + x_2 + \ldots + x_n$. Here $x_i \wedge x_j \leq x_j$. Since $x_1^2 < x_1$ we have a lesser expression for x. This is a contradiction. So $x^2 < x$ for all x. So some iterated square $m^{2^{n+1}}$ of m is zero, where m is the maximal element.

Suppose $m^{2^{n+1}} = 0$ but $m^{2^n} \neq 0$. We will construct a nonzero commutative distributive product such that $(x * y) * z = 0$ and $x * (y * z) = 0$ always which will therefore be associative. If $m^2 = 0$ then $xy = 0$. So assume $n > 0$.

Case 1. $m^{2^n} m^{2^{n-1}}$ or $m^{2^{n-1}} m^{2^n} > 0$. Suppose the the former. Let the product $x * y$ be $m^{2^n}(xy)$ which is nonzero for $x = y = m$. Then $(x * y) * z = m^{2^n}((x * y)z) = m^{2^n}((m^{2^n}xy)z)$ which is zero since $m^{2^n} m^{2^n}$ is an upper bound for it. Likewise $x * (y * z) = m^{2^n}(x(y * z)) = m^{2^n}(x(m^{2^n}(yz)) = 0$. This product is distributive, commutative, and $x * y \leq x$.

Case 2. Both those are zero. Let x * y be $x_1 y_1$
where x_1, y_1 are obtained by substituting x, y for som
two factors of $m^{2^{n-1}}$. (The case n = 1 is handled as
before.) Then x * (y * z) is bounded by $x_1(y_1 z_1)$ whic
in turn is bounded by $m^{2^{n-1}} m^{2^n} = 0$. Likewise (x * y)
* z = 0. □

PROPOSITION 1.2.8.   *Let W be the semilattice of
subspace of V where V is a vector space of dimension
at least 2 over a field. Then W has no incline structure except 0.*

Proof.   Any 1-dimensional subspace x is contained
in the span of two distinct 1-dimensional subspaces
y, z. So x + y = y + z = x + z, and xy = yz = xz = 0.
Then $y^2 = (y + z)(y + x) = (x + y)(x + z) = x^2$. So
$x^2 \leq \inf \{x, y\} = 0$. So any product of 1-dimensional
subspaces is zero.   □

The same proof shows that the lattice of partitions
of a set of at least three elements has no incline
structure since any basis element is contained in the
join of two others having meet zero.

Let Z_2 denote the 2-element field {0, 1} where
$0 + 0 = 1 + 1 = 1 \cdot 0 = 0 \cdot 1 = 0 \cdot 0 = 0, 1 \cdot 1 =$
$1 + 0 = 0 + 1 = 1$.

EXAMPLE 1.2.5.   The lattice of subspaces of $Z_2 \times Z_2$

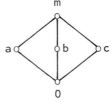

has no nontrivial incline structure.

EXAMPLE 1.2.6. The lattice of equivalence relations on a 3-element set is isomorphic to the previous example and has no nontrivial incline structure.

1.3 REPRESENTATIONS OF INCLINES

We shall start with the following definition.

DEFINITION 1.3.1. An incline with 0 is x-*simple* if and only if it has an element $x > 0$ such that any nontrivial congruence sends x to 0.

THEOREM 1.3.1. *There exists a 1-1 incline homomorphism from any finite incline 7 to a product of x-simple inclines.*

Proof. For any $y \neq z$ in 7, we will show there exists an x-simple quotient of 7 in which y, z have distinct images. Then the mapping into the product set will be 1-1.

Suppose by symmetry, that $y \not> z$. Take a congruence sending y and all elements less than y to 0. Then z will go to a nonzero element. Now take a maximal congruence on the result under which z does not go to 0. The result is an x-simple incline. □

This enables many questions of representation of inclines to be studied by means of x-simple inclines.

THEOREM 1.3.2. *A finite incline other than 0 is x-simple if and only if (i) x is the unique minimal nonzero element, (ii) for all distinct y, $z \neq 0$, there exists a such that $ay \geq x$, $az = 0$ or $ay = 0$, $az \geq x$.*

Proof.   Assume the incline is x-simple. We first show x is the unique minimal nonzero element. Suppose y $\not>$ x. Then we can take a congruence sending y and all elements less than or equal to y to 0, but not x. This is a contradiction.

Suppose we take a congruence y $\sim$ z for y, z distinct and nonzero. Suppose it identifies x with 0. We have a chain of relations ay + b $\sim$ az + b leading from 0 to x. Take the first one which sends from 0 to a nonzero element. Then ay + b = 0, az + b $\geq$ x. So b = 0. So ay = 0, az $\geq$ x.

Conversely, suppose (i) and (ii) hold. Any congruence sending y to 0 by (i) sends x to 0. By (ii) any congruence sending y to z if both are nonzero sends x to 0. $\square$

EXAMPLE 1.3.1.   The only x-simple distributive lattice is {0, 1}.

EXAMPLE 1.3.2.   Suppose we have an incline structure on

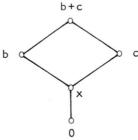

in which 0 $\cdot$ y = 0, x $\cdot$ y = 0 for all y and b^2 = x, c^2 = x, b $\cdot$ c = 0. This incline is x-simple.

DEFINITION 1.3.2.   An incline is *linearly (cyclically, finite) representable* if and only if there exists a 1-1 homomorphism from it into a product of inclines which are linearly ordered (cyclic, finite).

EXAMPLE 1.3.3.   A product of linearly ordered inclines is linearly representable.

EXAMPLE 1.3.4.   The incline generated by x, y subject to all relations $(x + y)^n = x^n + y^n$ is linearly representable. Mappings $x \to -n$, $y \to -m$ into Z give a set of linear representations. Here Z denotes the set of all integers.

Representing an incline linearly may simplify the proofs of various results about it. The most useful inclines may well be the linear ones on [0, 1] such as the *fuzzy algebra*.

THEOREM 1.3.3.   *If a finite incline 7 is linearly representable then for all a, b, y, z it satisfies $ay \wedge bz \le az + by$.*
   Proof.   Suppose 7 is linearly representable, by a mapping f. In any linear incline (therefore in a product of linear inclines) $ay \wedge bz \le az + by$ since by symmetry we may assume $y \le z$ so that $ay \le az$. Thus $f(ay \wedge bz) \le f(ay) \wedge f(bz) \le f(az + by)$. Since f is 1-1 the result holds.   □

Is the converse true ? For x-simple 7 it is true since we can define a complete binary relation R by a R b if $ac \ge x$, $bc = 0$ for some c. Then if $a \ge b$, we cannot have b R a so a R b. Conversely, suppose $ac \ge x$, $bc = 0$, but $a + b > a$. Let $(a + b)y = x$, $ay = 0$. Then $ac \wedge by \ge x > bc + ay$. This is a contradiction. So $a \ge b$ if and only if a R b. So $a \ge b$

is complete. However, an x-simple quotient of a linearly representable incline may not be linearly representable, as Example 1.3.5 below shows (it is a quotient of an incline 7 with generators a, b, c and rela-
tions $(x + y)^n = x^n + y^n$ and $x^r y^s z^t \leq x^{r+u} y^{s-u} z^t +$
$x^r y^{s+v} z^{t-v} + x^{r-w} y^s z^{t+w}$, for all x, y, z ε 7 and u, v,
w, r, s, t ε z^+. Here z^+ denotes the set of all positive integers. Any map from {a, b, c} to (0, 1 ,
sup {x, y}, xy) gives a representation).

Many other identities hold in linearly representable inclines, such as $(x + y)^n = x^n + y^n$.

EXAMPLE 1.3.5.   An incline which is distributive as
a lattice may not be linearly representable.

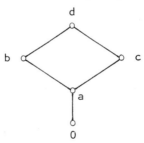

Let bc = 0, b^2 = b, c^2 = c, ad = 0. Then this is x-
simple but not linear.

In any linearly representable incline we have that
if ua $\leq$ vb and ub $\leq$ va then ua $\leq$ va and ub $\leq$ vb.
Assume by symmetry a $\leq$ b in a linear representation.
Then ua $\leq$ ub $\leq$ va and ub $\leq$ va $\leq$ vb. This result is
reminiscent of the *Thomsen Condition* in the theory of
additive conjoint measurement. The Thomsen Condition
gives a necessary condition for a complete quasiorder
on a Cartesian product **R** $\times$ **R** to be expressible by
$(x_1, y_1) \leq (x_2, y_2)$ if and only if $f(x_1) + g(y_1) \leq$

$f(x_2) + g(y_2)$. Here **R** denotes the set of all real numbers.

THOMSEN CONDITION.   *If* $(a, x) \geq (f, q)$ *and* (f, p) $\geq (b, x)$ *then* $(a, p) \geq (b, q)$.

However the Thomsen Condition itself does not hold for inclines since we could have $x = f = 0$, ap, bq any nonzero products. See Kim and Roush (1980).

PROPOSITION 1.3.4.   *A finitely generated incline* **7** *is finitely representable if and only if for all* $x \neq y$ ε **7**, *there exists n such that for the congruence* $\mathfrak{C}_n$ *identifying* z^n *with* z^{n+1} *for all* $z \varepsilon$ **7**, *x is not congruent to y.*

Proof.   It will suffice to show that the quotient inclines are finite, and that every finite incline satisfies a relation of this form. In any incline of order n the powers $x \geq x^2 \geq \ldots \geq$ are not distinct. So $x^r = x^{r+1}$ for some $r \leq n$. This implies $x^n = x^{n+1}$ for any x.

The quotient incline will be finitely generated and will satisfy $x^n = x^{n+1}$. Therefore, every element equals a polynomial having degree of most n in each generator. The number of such polynomials is finite. Hence the quotient incline is finite.   □

EXAMPLE 1.3.6.   The incline with generators x, y and no relations is finitely representable.

Finally, we summarize previous work on structures related to inclines: Cuninghame-Green's *belts* and *blogs* and Zimmerman's *idempotent semirings*.

R. Cuninghame-Green (1979) calls a commutative semi-ring satisfying x $\oplus$ x = x, a *belt*. He introduces a dual addition x $\oplus$' x = x (in practice $\oplus$ is often maximum and $\oplus$' is minimum) and a dual multiplication $\otimes$' which differs only in that $-\infty$ $\otimes$'∞ = $-\infty$ and $-\infty$ $\otimes$' ∞ = ∞ (in practice x $\otimes$ y is x + y). The combined structure is called a *blog*. He studies conjugacy, duality, residuation, special matrices all of whose row sums are equal, simultaneous linear equations, a definition of linear independence based on Σ $x_i v_i$ = c having unique solutions x_i for some c, rank based on this concept, regularity based on a strong linear independence, semi-norms, eigenvalues, eigenvectors, spectral inequalities sequences of powers, permanents, strict invertibility AB = BA = I. He then presents applications of these concepts.

U. Zimmerman (1981) deals with more general order structures: lattice ordered groups, idempotent semi-rings (the same as belts above), and ordered semimodules. He proves characterization and representation theorems for these. He then applies these to a simila set of problems to those considered by Cuninghame-Gree: shortest and other paths in graphs, linear programs, scheduling, duality for matroid flows, transportation and assignment problems. He also takes up eigenvalues in the finite dimensional case.

H. Hashimoto (1983) has studied what he calls a *path algebra*. A path algebra is a commutative semirin with 0, 1 satisfying x + x = x. Thus it does not have the incline inequality, so that many properties of inclines are false.

Our approach here is somewhat different: we conside algebraic ideas such as varieties, ascending chains, representation, Green's relational classes, regular elements defined by axa = a, series of powers of any matrix, generalized inverses. We prove eigenvalue theorems in the infinite dimensional case. Few of our theorems can be found in the other books and articles.

1.4 OPEN PROBLEMS

1. What linear inclines can be represented by the real numbers ?

2. Are finitely generated inclines finitely representable ?

3. What results of this chapter generalize to finitely generated inclines ? compact inclines ?

4. When will the lattice of normal subgroups of a group have a nontrivial incline structure ?

5. Is the condition of Theorem 1.3.3 necessary and sufficient ?

6. Is the following condition necessary and sufficient for linear representability of finite 7 ? Let y_i $\varepsilon\ 7$ for $i = 1$ to n and let M_i be monomials such that M_i involves y_i. Let N_i be obtained from M_i by changing one or more y_i factors to y_{i+1} (y_n to y_1 for M_n). Then

$$\wedge\ M_i\ \leq\ \vee\ N_i$$

Chapter 2

Vectors and Matrices
over an Incline

In this chapter we present the theory of vectors and
matrices over an incline. Matrices over an incline
form a semiring, and many results not involving inverses can be taken from fields. For many incline only
permutation matrices are invertible. Subspaces of
vectors have bases. There exists a special kind of
basis, a standard basis, which is unique if idempotents
of the incline are linearly ordered. Green's relations
can be characterized in terms of the spaces of vectors
spanned by the rows (columns) of a matrix.

2.1 DEFINITION OF OPERATIONS

Let V_n for an incline 7 denote the Cartesian product
incline $7 \times 7 \times \ldots \times 7$ (n factors). Then V_n is called
the set of n-*dimensional vectors over the incline*. It
is itself an incline. The set of n × m matrices over
an incline 7 is also an incline under operations: (i)
(i) $A + B = (a_{ij} + b_{ij})$; (ii) $A \odot B = (a_{ij}b_{ij})$.

For brevity, let $M_{nm}(7)$ denote the set of all $n \times m$ matrices over an incline 7. Matrix multiplication can be defined by $AB = \sum\limits_{k} a_{ik}b_{kj}$.

EXAMPLE 2.1.1. In the incline $[0, 1]$ with operations sup $\{A, B\}$ and AB:

$$\begin{bmatrix} 0.1 & 0.3 \\ 0.5 & 0.2 \end{bmatrix} + \begin{bmatrix} 0.1 & 0.4 \\ 0.4 & 0.1 \end{bmatrix} = \begin{bmatrix} 0.1 & 0.4 \\ 0.5 & 0.2 \end{bmatrix}$$

$$\begin{bmatrix} 0.1 & 0.3 \\ 0.5 & 0.2 \end{bmatrix} \odot \begin{bmatrix} 0.1 & 0.4 \\ 0.4 & 0.1 \end{bmatrix} = \begin{bmatrix} 0.01 & 0.12 \\ 0.20 & 0.02 \end{bmatrix}$$

$$\begin{bmatrix} 0.1 & 0.3 \\ 0.5 & 0.2 \end{bmatrix} \times \begin{bmatrix} 0.1 & 0.4 \\ 0.4 & 0.1 \end{bmatrix} = \begin{bmatrix} 0.12 & 0.04 \\ 0.08 & 0.20 \end{bmatrix}$$

As with any semiring, we have these laws: (i) Commutative) $A + B = B + A$; (ii)(Associative) $(A + B) + C = A + (B + C)$, $(AB)C = A(BC)$; (iii) (Distributive) $A(B + C) = AB + AC$, $(B + C)A = BA + CA$. In addition $A + A = A$. *Inequality* of matrices is defined by $A \le B$ if and only if $a_{ij} \le b_{ij}$ for all i, j. *Transpose* is defined by $A^T = (a_{ji})$.

EXAMPLE 2.1.2.

$$\begin{bmatrix} 0.1 & 0.4 \\ 0.7 & 0.5 \end{bmatrix} < \begin{bmatrix} 0.3 & 0.4 \\ 0.8 & 0.6 \end{bmatrix}$$

$$\begin{bmatrix} 0.1 & 0.4 \\ 0.5 & 0.5 \end{bmatrix}^T = \begin{bmatrix} 0.1 & 0.5 \\ 0.4 & 0.5 \end{bmatrix}$$

To illustrate the above proofs, we give the proof of a distributive law.

$$A(B + C) = \left(\sum_k a_{ik}(b_{kj} + c_{kj})\right) = \left(\sum_k a_{ik}b_{kj}\right)$$

$$+ \left(\sum a_{ik}c_{kj}\right) = AB + AC$$

The relation "$\leq$" (less than or equal) makes the semiring of matrices over any ordered semiring into an ordered semiring, that is, if $A \leq B$ then $AC \leq BC$, $CA \leq CB$, $A + C \leq B + C$ for all C.

Vectors can be regarded as $1 \times n$ matrices (*row vectors*) or $n \times 1$ (*column vectors*). This gives the laws for multiplying vectors and matrices. There exists a scalar multiplication for $c \, \epsilon \, 7$: $cA = (ca_{ij})$. It satisfies: (i) $b(cA) = (bc)A$; (ii) $(b + c)A = bA + cA$; (iii) $c(AB) = (cA)B = A(cB)$, and (iv) $c(A + B) = cA + cB$. Vectors have a scalar product vw^T.

EXAMPLE 2.1.3.

$$0.2 \begin{bmatrix} 1 & 0 \\ 0 & 0.5 \end{bmatrix} = \begin{bmatrix} 0.2 & 0 \\ 0 & 0.1 \end{bmatrix}$$

$$\begin{bmatrix} 0.1 & 0.2 \end{bmatrix} \begin{bmatrix} 0.1 \\ 0.2 \end{bmatrix} = 0.04$$

In this chapter we assume henceforth that an additive identity 0 and a multiplicative identity 1 exist in 7. It follows that 0 is the lowest element, and 1 is the highest element of 7. And $0x = 0$, $1 + x = 1$.

DEFINITION 2.1.1. The $n \times m$ *zero matrix* is $0 = (o_{ij})$ where $o_{ij} = 0$ for all i, j. The $n \times m$ *identity matrix* is $I = (\delta_{ij})$ where $\delta_{ij} = 1$ if $i = j$ and $\delta_{ij} = 0$ otherwise. The *universal matrix* J is the matrix all of whose entries are 1.

EXAMPLE 2.1.4.

$$\begin{bmatrix} 0 & 0 \\ 0 & 0 \end{bmatrix}, \quad \begin{bmatrix} 1 & 0 \\ 0 & 1 \end{bmatrix}, \quad \begin{bmatrix} 1 & 1 \\ 1 & 1 \end{bmatrix}$$

We have for any A, $0 + A = A$, $0A = 0$, $AI = IA = A$.
We next show that few elements of an incline have
inverses.

DEFINITION 2.1.2.   A matrix P is a *permutation*
matrix if and only if its entries are 0, 1 and it has
exactly one 1 in each row and column.

EXAMPLE 2.1.5.

$$\begin{bmatrix} 0 & 1 & 0 \\ 0 & 0 & 1 \\ 1 & 0 & 0 \end{bmatrix}$$

This condition means the set of ordered pairs (i, j)
with $p_{ij} = 1$ form a permutation π.

DEFINITION 2.1.3.   An incline $\mathcal{I}$ with 0, 1 is *inte-*
gral if and only if there do not exist nonzero x, y ε $\mathcal{I}$
such that $xy = 0$, $x + y = 1$.   (This generalizes the
previous definition of Cao.)

PROPOSITION 2.1.1.   *Every finite incline is a direct*
sum of integral inclines.
    Proof.   Let $xy = 0$, $x + y = 1$.   We have $x \geq x^2 \geq \cdots$
and $y \geq y^2 \geq \cdots$.   Choose n such that $x^n = x^m$ and
$y^n = y^m$ for $m > n$.   Then $x^n y^n = 0$ and $1 = (x + y)^n =$
$x^n + y^n$.   Also x^n, y^n are idempotent.

Moreover, $7x^n$ and $7y^n$ are subinclines. There is a mapping $7x^n \oplus 7y^n \to 7$ given by $(a, b) \to a + b$. This is a incline homomorphism since $x^n y^n = 0$. There is an inverse given by $c \to (cx^n, cy^n)$. So 7 is a direct sum. Repeat this process as far as possible. The remaining summands must be integral. ☐

2.2 BASIS

We start with the following lemma.

LEMMA 2.2.1. *In an integral incline, if* $a_1 + a_2 + \ldots + a_n = 1$, $a_i a_j = 0$ *for* $i \neq j$ *then at least one* a_i *is 1 and the rest are zero.*

Proof. $a_1 + (a_2 + \ldots + a_n) = 1$ and $a_1(a_2 + \ldots + a_n) = 0$. Thus $a_1 = 0$ or $a_2 + a_3 + \ldots + a_n = 0$. So some $a_i = 0$. Delete it. This reduces the case for n to the case for n - 1. For n = 1, the result is true. ☐

PROPOSITION 2.2.2. *For a finite incline let* $a^n = a^{n+1}$ *for every n. The mapping* $a \to a^n$ *is an incline homomorphism* $f: 7 \to S$ *where S is the lattice of idempotents.*

Proof. We have $f(x + y) = (x + y)^{2n} \leq x^n + y^n$ since each term of $(x + y)^{2n}$ is divisible by x^n or y^n. But $(x + y)^n \geq x^n + y^n$ so equality holds. And $(xy)^n = x^n y^n$. ☐

DEFINITION 2.2.1. A semiring with 0, 1 is *positive* if and only if whenever $x_1 + x_2 + \ldots + x_n = 0$ then $x_i = 0$ for each i.

EXAMPLE 2.2.1. Any incline is positive.

EXAMPLE 2.2.2. For any n, $(Z^+)^n$ is positive.

THEOREM 2.2.3. *In any commutative positive semiring if $XY = I$ then (i) for $k \neq j$, $x_{ij}x_{ik} = x_{ji}x_{ki} = 0$, $y_{ij}y_{ik} = y_{ji}y_{ki} = 0$, (ii) $x_{ik}y_{kj} = x_{ki}y_{jk} = 0$ for $i \neq j$, (iii) the product of the ith row sum of X and the jth column sum of Y equals 1, as does the product of the jth column sum of X and the ith row sum of Y, (iv) $YX = I$, (v) the semiring splits as a direct sum such that in each nonzero part there exists a permutation P such that if $j \neq p(i)$, $x_{ij} = y_{ji} = 0$ and if $j = p(i)$, $y_{ji} = x_{ij}^{-1}$.*

Proof. Since $XY = I$, $\Sigma\, x_{ik}y_{kj} = \delta_{ij}$. So for $i \neq j$, $x_{ik}y_{kj} = 0$. Expand $\prod_i \left(\Sigma_k\, x_{ik}y_{kj} \right) = \prod 1 = 1$. All terms $x_{ik}y_{ki}x_{jk}y_{kj}$ are zero for $i \neq j$ since $x_{ik}y_{kj} = 0$. So this product equals $\Sigma_p \prod x_{ip(i)}y_{p(i)i}$ when p ranges over all permutations. Now for $j \neq k$,

$$x_{ik}x_{jk} = x_{ik}x_{jk} \,\Sigma_p \prod x_{ip(i)}y_{p(i)i} = 0$$

since in each summand $x_{ik}x_{jk} \prod x_{ip(i)}y_{p(i)i}$, there occurs a factor $x_{ik}x_{jk}y_{kq(k)}$ ($q(k) = p^{-1}(k)$ and either $q(k) \neq i$ or $q(k) \neq j$). By a similar reasoning each term of $x_{ki}x_{kj} \Sigma_p \prod x_{ip(i)}y_{p(i)}$ contains a factor $x_{ki}x_{kj}y_{iq(i)}y_{jq(j)}$ which is 0 since either $k \neq q(i)$ or $k \neq q(j)$. A similar argument holds for y. This proves (i).

We have already proved the first equation of (ii). The second part follows from the equations

$$x_{ki}y_{jk} = x_{ki}y_{jk} \,\Sigma_p \prod x_{ip(i)}y_{p(i)i} = 0$$

since each term has a factor $x_{ki}y_{jk}y_{iq(i)}x_{q(j)j} = x_{ki}y_{iq(i)}x_{q(j)j}y_{jk}$ and either $k \neq q(i)$ or $q(j) \neq k$. This proves (ii).

Now $\Sigma_k\, x_{ik} \Sigma_m\, y_{mi} = \Sigma_k\, x_{ik}y_{ki} = 1$. The quantities

$e_{ik} = x_{ik}y_{ki}$ are idempotents since

$$(x_{ik}y_{ki})^2 = x_{ik}y_{ki} \sum_k x_{ik}y_{ki} = x_{ik}y_{ki}(1) = x_{ik}y_{ki}$$

For any permutation P, define $e_p = \prod_i e_{ip(i)}$. Then e_p

is an idempotent. We have $\sum_p e_p = \sum_p \prod_i x_{ip(i)}y_{p(i)i} = 1$.

For $p \neq r$, $e_p e_r = 0$ since there exists $i \neq j$ such that $p(i) = r(j)$ and the product $\prod x_{ip(i)}y_{p(i)i}x_{ir(i)}y_{r(i)i}$ has a factor $x_{ip(i)}y_{p(i)j} = 0$. Therefore, the e_p form a system of orthogonal idempotents.

Therefore, the semiring **SR** breaks into the direct sum of the semirings e_p**SR** over all permutations P such that $e_p \neq 0$. If $e_p \neq 0$ then $e_p X e_p Y = e_p I$ so $e_p X$, $e_p Y$ are nonzero. We have $e_p x_{ij} = 0$ unless $j = p(i)$ since there is a factor $x_{ij}y_{j\tau(j)}$, $\tau(i) = p^T(i)$. And $e_p y_{ji} = 0$ for $j \neq p(i)$ imply $e_p Y e_p X = e_p I$, as well as (v). So $YX = I$. This implies the rest of (iii). □

For simplicity, R_0^+ (Z_0^+) will denote the nonnegative real numbers (integers).

EXAMPLE 2.2.3.   Over $R_0^+ \times R_0^+$

$$I = \begin{bmatrix} (0,\,0) & (1,\,0) & (0,\,\frac{1}{4}) \\ (0,\,\frac{1}{2}) & (0,\,0) & (1,\,0) \\ (1,\,0) & (0,\,\frac{1}{3}) & (0,\,0) \end{bmatrix} \begin{bmatrix} (0,\,0) & (0,\,2) & (1,\,0) \\ (1,\,0) & (0,\,0) & (0,\,3) \\ (0,\,4) & (1,\,0) & (0,\,0) \end{bmatrix}$$

COROLLARY 2.2.4.   *In an incline if* $XY = I$ *then* $Y = X^T$ *and* $YX = I$.

COROLLARY 2.2.5.   *In an integral incline if* $XY = I$ *then* X, Y *are permutation matrices.*

DEFINITION 2.2.2.   A set W of vectors is a *subspace* if and only if for all a ϵ 7, u, v ϵ W, av ϵ W, and u + v ϵ W.

EXAMPLE 2.2.4.   The set V_n is a subspace. For any c ϵ 7, cV_n is a subspace.

DEFINITION 2.2.3.   A vector v is *linearly dependent* on a set S if $v = \Sigma a_i w_i$, w_i ϵ S. A set S of vectors is *linearly independent* if v is not linearly dependent on S \ {v} for any v ϵ S.

EXAMPLE 2.2.5.   The vectors $(\frac{1}{2}, \frac{1}{4})$, $(1, \frac{1}{2})$ are linearly dependent over the incline (0, 1] with standard multiplication.

DEFINITION 2.2.4.   The *span* of a set W of vectors is the set <W> of all linear combinations $\Sigma a_i w_i$, w_i ϵ W.

EXAMPLE 2.2.6.   The span of {(0, 1), (1, 0)} is all of V_2.

PROPOSITION 2.2.6.   *The span of any subset is a subspace.*

  Proof. $c(\Sigma a_i w_i) = \Sigma c a_i w_i$

          $(\Sigma a_i w_i) + (\Sigma b_j w_j) = \Sigma a_i w_i + \Sigma b_j w_j$ $\square$

DEFINITION 2.2.5.   A *basis* for a subspace W of V_n is a linearly independent set S of vectors such that <S> = W. An *i-basis* is a basis which is a subset of a lattice basis for W. A *standard basis* is one {x_1, x_2, ... , x_n} such that if $x_i = \Sigma c_{ij} x_j$ then $x_i = c_{ii} x_i$.

DEFINITION 2.2.6.   The *row* (*column*) *space* for a
matrix is the subspace spanned by its row (column)
vectors. The row (column) space of a matrix A is de
noted by R(A) (C(A)). A *row* (*column*) *basis* $B_r(A)$
($B_c(A)$) of a matrix A is an basis for the row (column)
space. The *row* (*column*) *rank* $\rho_r(A)$ ($\rho_c(A)$) of a matrix
A is the cardinality of an i-basis for the row (column)
space.

EXAMPLE 2.2.7.   The row space R(I) of an identity
matrix I is all of V_n. A row basis is formed by its
rows. So $\rho_r(I) = n$, provided that 1 is a lattice basis
element of 7.

EXAMPLE 2.2.8.   Over the fuzzy algebra {(0.5, 1),
(0.6, 0.7)} is a basis for the set it spans, but not a
standard basis since (0.6, 0.7) = (0.6, 0.6) + (0.5,
0.7) is not a lattice basis element.

For brevity, we will denote the ith row (column) of
a matrix A as A_{i*} (A_{*i})

PROPOSITION 2.2.7.   *The ith row of XY is* $\Sigma\ x_{ik}Y_{k*}$.
Its ith column is $\Sigma\ X_{*k}y_{kj}$.
   Proof.   The jth-entry of $\Sigma\ x_{ik}Y_{k*} = \Sigma\ x_{ik}Y_{kj}$ which
is the (i, j)-entry of XY.   □

This implies that R(XY) ⊂ R(Y) and is the subspace
spanned by $\Sigma\ x_{ik}Y_{k*}$. There exists an epimorphism from
R(X) to R(XY) sending v to vY.
   It is possible for a 1 × 1 matrix to have rank 2.

EXAMPLE 2.2.9.   Over the Boolean algebra {0, 1} ×

$\{0, 1\}$, the matrix $[1 \quad 1]$ has rank 2 since a row basis is $\{(0, 1), (1, 0)\}$.

Proposition 2.2.7 also holds for infinite compact subspaces whenever a lattice basis exists.

For brevity, let $\underline{n} = \{1, 2, \ldots, n\}$.

PROPOSITION 2.2.8. *Every compact subsemilattice of $[0, 1]^n$ has a semilattice basis.*

Proof. For any $i \in \underline{n}$ and any $\alpha \in (0, 1]$, let $B_{i\alpha}$ be the set of minimal elements y such that $y_i = \alpha$ if that set is nonempty. Then $\cup \, B_{i\alpha}$ is independent. Suppose $y = \Sigma \, w\langle j \rangle$. Then some $w\langle j \rangle_i = \alpha$, and $w\langle j \rangle \leq y$. So $w\langle j \rangle = y$ by minimality. They span the subsemilattice. For any x and all i choose some $y\langle i \rangle$ such that $y_i\langle i \rangle = x_i$ and $y_i \leq x$. Then $x = \Sigma \, y\langle i \rangle$. $\square$

2.3 GREEN'S RELATIONS

In this section we discuss Green's equivalence classes (Clifford and Preston (1961)). As usual, here L, R, H, D, J stand for Green's relations.

THEOREM 2.3.1. *For matrices over any semiring: (i) $A \, L \, B$ if and only if $R(A) = R(B)$, (ii) $A \, R \, B$ if and only if $C(A) = C(B)$, (iii) $A \, H \, B$ if and only if $R(A) = R(B)$ and $C(A) = C(B)$, (iv) $A \, D \, B$ if and only if there exist matrices X, Y such that X gives an isomorphism $R(A)$ to $R(B)$ and Y gives the inverse mapping $R(B)$ to $R(A)$, (v) $A \, J \, B$ if and only if there exist matrices X, Y such that X gives a homomorphism from an n-generator subspace of $R(A)$ onto $R(B)$ and Y gives a homomorphism from an n-generator subspace of $R(B)$ onto $R(A)$.*

Proof.   (i) If XA = B and YB = A then $R(A) \subset R(B)$ and $R(B) \subset R(A)$ so $R(B) = R(A)$. Let $R(A) = R(B)$. Then there exist x_{ik} such that $B_{i*} = \Sigma \ x_{ik}A_{k*}$. Then B = XA. Likewise A = YB.

The proof of (ii) is similar. From (i) and (ii), the statement (iii) follows.

Let A $\mathcal{D}$ B. Then there exists X such that A $\mathcal{R}$ AX $\mathcal{L}$ B, and Y such that AXY = A. Then R(AX) = R(B). So X gives an isomorphism from R(A) to R(AX) = B and Y its inverse. Conversely, if X, Y are as in the statement of the theorem, then A $\mathcal{R}$ AX and AX has the same row space as B, hence AX $\mathcal{L}$ B.

Let B = XAY. Then R(XA) is an n-generator subspace of R(A) and Y maps it onto R(B). Conversely, any n-generator subspace of R(A) has the form R(XA) where the rows of XA are the generators. If Y gives a homomorphism onto R(B) then R(XAY) = R(B) so XAY, B are $\mathcal{L}$-equivalent. So UXAY = B for some U. This implies the last statement. $\square$

EXAMPLE 2.3.1.

$$\begin{bmatrix} 1 & 1 \\ 1 & 1 \end{bmatrix} \ \mathcal{L} \ \begin{bmatrix} 1 & 1 \\ 0 & 0 \end{bmatrix}, \qquad \begin{bmatrix} 1 & 1 \\ 1 & 1 \end{bmatrix} \ \mathcal{R} \ \begin{bmatrix} 1 & 0 \\ 1 & 0 \end{bmatrix}$$

$$\begin{bmatrix} 1 & 1 \\ 1 & 1 \end{bmatrix} \ \mathcal{D} \ \begin{bmatrix} 1 & 0 \\ 0 & 0 \end{bmatrix}, \qquad \begin{bmatrix} 0 & 1 & 0 \\ 0 & 0 & 1 \\ 1 & 0 & 0 \end{bmatrix} \ \mathcal{H} \ \begin{bmatrix} 1 & 0 & 0 \\ 0 & 1 & 0 \\ 0 & 0 & 1 \end{bmatrix}$$

DEFINITION 2.3.1.   An incline is *metric* if there exists a function d(x, y) on a set S to R_0^+ such that (i) d(x, y) = 0 if and only if x = y, (ii) d(x, y) = d(y, x), (iii) d(x, y) + d(y, z) $\geq$ d(x, z), (iv) if x $\leq$ y $\leq$ z then sup {d(x, y), d(y, z)} $\geq$ d(x, z), (v) the operations (+, ·) are continuous in the metric d.

PROPOSITION 2.3.2.   *Suppose incline 7 is compact metric and for all $a \in 7$ either $a^2 = a^3$ or $a^n \to 0$. Then $J = D$ for finite matrices over 7.*

Proof.   Let $XAY = B$, $ZBW = A$. Let 7_1 be the sub-incline generated by A, B, X, Y, Z, W. Choose $\varepsilon > 0$. Let n be large enough that for each generator a of 7_1 either $a^n = a^{n+1}$ or $d(a^n, 0) < \varepsilon$. *The quotient obtained by sending all points within ε of 0 to 0 is finite since every product of generators can be represented by a monomial of degree less than or equal to n in each generator, or 0.   Thus, in it we have* X_ε, Y_ε, Z_ε, W_ε *such that* $AX_\varepsilon Y = A$, $Z_\varepsilon A X_\varepsilon = B$, $Z_\varepsilon W_\varepsilon B = B$. Since 7 is compact, a subsequence converges to some points X_0, Y_0, Z_0, W_0. Then $AX_0Y_0 = A$, $Z_0AX_0 = B$, $Z_0W_0B = B$ in each quotient and therefore in 7. So $A\ D\ B$.   □

2.4 STANDARD BASES

Standard bases have a number of special properties especially in the fuzzy case, when they are unique. As before, V_n will denote vectors over 7 of dimension n.

PROPOSITION 2.4.1.   *For any finite incline, for any basis B there exists a standard basis spanning the same set, of the same cardinality as B.*

Proof.   If we have a basis which is not standard let $x_k = \Sigma\ c_i x_i$. If $c_k x_k < x_k$ replace x_k by $c_k x_k$. This set spans the same space since $x_k = \Sigma\ c_i x_i$.   No other x_j will now be be dependent. If $c_k x_k$ were dependent, then x_k would be since $x_k = \Sigma\ c_i x_i$.

This step decreases the basis. The process must terminate after a finite number of steps, since the incline and so V_n is finite.   □

LEMMA 2.4.2. *In a vector space over a finite incline, let $x = \Sigma\, c_i x$. Then for some n, $x = \Sigma\, c_i{}^n x$ where $c_i{}^n$ are idempotents.*

Proof. Since

$$x = \left(\sum_{i=1}^{k} c_i \right) x$$

we have

$$x = \left(\sum_{i=1}^{k} c_i \right)^{kn} x$$

In

$$\left(\sum_{i=1}^{k} c_i \right)^{kn}$$

each term involves some c_i to at least an nth power so

$$\left(\sum_{i=1}^{k} c_i \right)^{kn} x \geq \left(\sum_{i=1}^{k} c_i{}^n x \right)$$

So

$$x = \sum_{i=1}^{k} c_i{}^n x$$

Take n large enough that each $c_i{}^n$ is idempotent (since $c_i \geq c_i{}^2 \geq \ldots$, such n exists). $\square$

LEMMA 2.4.3. *Let x be a member of a standard basis over a finite incline and suppose $x = \Sigma\, y_i$ for y_i in the space spanned by the standard basis. If the idempotents in the incline are linearly ordered then $x = y_k$ for some k.*

Proof. Let $x = x_1$, and the basis be $x_1, x_2, \ldots, x_k$. Write $y_i = \Sigma\, c_{ij} x_j$. Then $x = \Sigma_j \left(\Sigma_i c_{ij} \right) x_j$. Thus $x = \Sigma\, c_{i1}{}^n x$, where $c_{i1}{}^n$ are idempotents. Let $c_{j1}{}^n$ be the largest. Then $x = c_{j1}{}^n x$. So $x \geq y_j \geq c_{j1} x \geq c_{j1}{}^n x$. Since equality holds, $x = y_j$. $\square$

This result enables us to generalize the result for fuzzy matrices to all finite inclines in which the idempotents are linearly ordered.

THEOREM 2.4.4.   (Cao). *For a finite incline in which idempotents are linearly ordered every subspace has a unique standard basis.*

Proof.   Let x_1, x_2, ... , x_k and y_1, y_2, ... , y_j be distinct standard bases. For any x_i we have $x_i = \Sigma c_{ik} y_k$ for some c_{ij}. So by Lemma 2.4.3, $x_i = c_{ik} y_k$ for some k. Conversely, $y_k = d_{kj} x_j$ for some j. If $i \neq j$ were possible, then x_i would be dependent. So $x_i = c_{ik} y_k = c_{ik} d_{ki} x_i$ so $c_{ik}^n d_{kj}^n x_i = x_i$. This shows the bases are equal.   □

EXAMPLE 2.4.1.   The assumption on idempotents is necessary else $(e_1 + e_2, 0)$ and $(e_1, 0)$, $(e_2, 0)$ are both standard bases for idempotents e_1, e_2.

COROLLARY 2.4.5.   *The cardinality of a basis is unique in the case described above.*

When does a subspace of V_n have a unique basis ? For integral inclines the theorem on inverses shows that V_n itself has a unique basis.

PROPOSITION 2.4.6.   *A basis B for a finite incline* 7 *is unique if and only if for all* $x \in B$, $z \notin B$, $B \setminus \{x\}$ *∪ $\{z\}$ does not contain a basis, and for all* $x \in B$ *if* $x = c_1 x + c_2 x$ *then* $x = c_1 x$ *or* $x = c_2 x$.

Proof.   The first condition is evidently necessary. If $x = c_1 x + c_2 x$ then $B \setminus \{x\}$ ∪ $\{c_1 x, c_2 x\}$ is a spanning set from which a subset will be a basis. Let $B = \{x_1, x_2, ... , x_n\}$ and suppose $B \setminus \{x_i\}$ ∪ $\{z\}$ never

contains a basis, and if $x = c_1 x + c_2 x$ then $x = c_1 x$
or $c_2 x$. So then B must be a standard basis else we
could replace x by cx for some x to obtain a new basis,
where $x = cx + \Sigma c_j x_j$ where $x_j \neq x$.

Now suppose $\{v_1, v_2, \ldots, v_n\}$ is another basis.
Suppose it does not contain $x \in B$. Write $v_i = c_i x + w_i$ where w_i is a sum of basis elements $x_i \neq x$. Then
$x = \Sigma d_j c_j x + \Sigma d_j w_i$. By the standard basis property,
$x = \Sigma d_j c_j x$. By the property about $x = c_1 x + c_2 x$
(extended by induction to n summands) $x = d_j c_j x$ for
some x. Thus $x = c_j x$. Since $x \geq d_j v_j$ we have $x = d_j v_j$.
Thus $B \setminus \{x\} \cup \{v_j\}$ is a spanning set and so contains
a basis.   □

EXAMPLE 2.4.2.   In the incline $\{0, \alpha, \beta, \alpha + \beta = 1\}$
where $\alpha\beta = 0$, $(1, 0)$ is not a unique basis since $(1, 0)$
$= \alpha(1, 0) + \beta(1, 0)$.

PROPOSITION 2.4.7.   *Suppose X L Y and the rows of
X form a unique basis for its row space.   Then for a
permutation P, PX = Y.*

Proof.   The rows of Y contain a basis so they must
coincide with the rows of X up to a permutation.   So
PX = Y.   □

EXAMPLE 2.4.3.

$$\begin{bmatrix} 0 & 1 & 1 \\ 1 & 0 & 1 \\ 1 & 1 & 0 \end{bmatrix} = \begin{bmatrix} 0 & 1 & 0 \\ 0 & 0 & 1 \\ 1 & 0 & 0 \end{bmatrix} \begin{bmatrix} 1 & 1 & 0 \\ 0 & 1 & 1 \\ 1 & 0 & 1 \end{bmatrix}$$

2.5 OPEN PROBLEMS

1. Is $J = D$ for all finitely generated inclines ?
    all compact inclines ?

2. Characterize all standard bases for a given i-basis

Chapter 3

Regularity and Inverses

Matrices, called *idempotent*, which equal their own squares are of special importance. Over the real numbers such a matrix is a projection onto a subspace. Over the Boolean algebra β they are related to partial orders. We show how to reduce idempotents so that the nonzero rows (or columns) form a standard basis. A certain triangular form also exists for idempotents. A matrix is called *regular* if it lies in a D-class of an idempotent. The only general method we have found for testing regularity and finding a generalized inverse is by computing the greatest inverse. However, if the incline is linearly ordered we can adapt algorithms for fuzzy inverses. We show that if a Moore-Penrose inverse of a matrix over an incline exists, it equals the transpose.

3.1 BOOLEAN IDEMPOTENTS

Over Boolean algebra $β = \{0, 1\}$, idempotent matrices have a particularly simple form. Deletion of dependent rows and columns does not affect idempotency and after we delete them we obtain an $r \times r$ idempotent of rank r, called a *nonsingular matrix*. Such idempotents E are reflexive and there exist permutation matrices P such that PEP^T is in triangular form with zeros above the main diagonal. For proofs, see Kim (1982).

PROPOSITION 3.1.1.   *A matrix M over any incline* 7 *is idempotent if and only if* $M_{i*} = \Sigma\ m_{ij}M_{j*}$.

Proof.   For any product BC, the ith row of the product equals $\Sigma\ b_{ij}C_{j*}$. $\square$

DEFINITION 3.1.1. 7_3 is the incline $[0, 1]$ under the operations sup $\{x, y\}$, and xy, 7_2 is $[0, 1]$, inf $\{x, y\}$, and inf $\{x + y, 1\}$. We will call the fuzzy algebra 7_1.

We give examples of idempotent matrices.

For brevity, let $M_n(\beta)$ denote the set of all $n \times n$ matrices over Boolean algebra $\beta = \{0, 1\}$.

EXAMPLE 3.1.1.   These are idempotents in $M_3(\beta)$.

$$\begin{bmatrix} 1 & 0 & 0 \\ 0 & 1 & 0 \\ 1 & 1 & 0 \end{bmatrix}, \quad \begin{bmatrix} 1 & 0 & 0 \\ 1 & 1 & 0 \\ 1 & 1 & 1 \end{bmatrix}, \quad \begin{bmatrix} 0 & 0 & 0 \\ 1 & 1 & 0 \\ 1 & 1 & 0 \end{bmatrix}$$

EXAMPLE 3.1.2.   Over a general incline we cannot assume idempotents have main diagonal entries which are idempotent. Let $0 < x < 1$ and $7 = 7_3 \times \beta$

$$\begin{bmatrix} (1, 0) & (x, 0) \\ (x, 0) & (x^2, 1) \end{bmatrix}$$

EXAMPLE 3.1.3.   Over 7_1 we cannot always delete dependent rows in a unique way. Suppose we have

$$\begin{bmatrix} 1 & 0 & 0 \\ 1 & 0.5 & 0.5 \\ 0.5 & 1 & 0.5 \\ 0 & 1 & 0 \end{bmatrix}$$

Then A_{2*} is dependent on A_{3*} and A_{1*}, A_{3*} is dependent on A_{2*} and A_{4*}. Yet if we delete A_{2*}, A_{3*} we have a smaller row space.

3.2 REDUCTION OF IDEMPOTENTS OVER 7

Although we cannot replace dependent rows in one step in an idempotent over 7 by zero, we can diminish the rows in steps until all dependent rows are zero and the nonzero rows form a standard basis. Idempotents with this property have idempotent main diagonal entries.

THEOREM 3.2.1.   *In any incline* 7, *for any positive integer* t *if A is idempotent in* $M_n(7)$, *then* $a_{ij} = \Sigma\ a_{ik} a_{kk}{}^t\ a_{kj}$. *Here* $M_n(7)$ *stands for the set of all* $n \times n$ *matrices over* 7.

Proof.   We have $a_{ij} \geq a_{ik} a_{kk}{}^t a_{kj}$ since the $a_{ik} a_{kk}{}^t a_{kj}$ is a term in the expansion of $a_{ij}{}^{(t+2)}$ Expand $a_{ij}{}^{(s)}$ where $s > tn^2 + n + 1$. Each term is a product $a_{ii_1} a_{i_1 i_2} \cdots a_{i_{s-1}j}$. Among the i_u some number k occurs at least $nt + 1$ times since there are $s - 1 > n^2 t + n$ terms in the sequence and only n distinct values for the i_j. Then this term is less than or equal to $a_{ik} a_{kk} \cdots a_{kk} a_{kj}$ where a_{kk} occurs t times, since for example $a_{kk} \geq a_{kj_1} \geq \cdots \geq a_{j_w k}$ for any

sequence $j_1 j_2 \cdots j_w$. Thus each term of $a_{ij} = a_{ij}^{(s)}$
is less than or equal to a term $a_{ik} a_{kk}^{t} a_{kj}$. So a_{ij}
$\leq \Sigma \, a_{ik} a_{kk}^{t} a_{kj}$. $\square$

DEFINITION 3.2.1.   An incline is *Type I* if it has
a Hausdorff topology in which products are continuous
and the powers x^m of each element x have a limit and
the distributive lattice of idempotents has a basis.

EXAMPLE 3.2.1.   Any finite incline is Type I.

PROPOSITION 3.2.2.   *Let 7 be Type I. Then if A is*
idempotent in $M_n(7)$, $a_{ik} = \Sigma \, a_{ik} e_{kk} a_{kj}$ *for all i, j*
where e_{kk} *is the limit of* a_{kk}^{n}.
     Proof.   This follows immediately from the previous
result. $\square$

EXAMPLE 3.2.2.   The same basis can yield two different standard bases of its multiples. Over β^3 the
basis $\{(1\ 1\ 0), (0\ 1\ 1)\}$ yields $\{(1\ 1\ 0), (0\ 0\ 1)\}$
and $\{(1\ 0\ 0), (0\ 1\ 1)\}$, which are both standard. Here
$\beta^3 = \beta \times \beta \times \beta$

EXAMPLE 3.2.3.   It is not always possible to delete
dependent rows in an idempotent over a general incline.
Let x, y be a basis for the 4-element Boolean algebra.
Suppose
$$\begin{bmatrix} x & 0 & x \\ 0 & 0 & y \\ x & 0 & 1 \end{bmatrix}$$

Then $A_{3*} = A_{1*} + A_{2*}$ but A_{3*} cannot be deleted.

THEOREM 3.2.3.   *Let* E *be any idempotent over finite*
7. Then there exists an idempotent $F \leq E$ *such that*
(i) $R(F) = R(E)$, *(ii) the nonzero rows of* F *are a stan-*
dard basis, (iii) $f_{ii}F_{i*} = F_{i*}$ *for all* i, *(iv)* f_{ii} *is*
idempotent for all i.

Proof.   Suppose first that $E_{i*} \neq e_{ii}E_{i*}$. For any
idempotent we have $E_{j*} = \Sigma\ e_{jk}E_{k*}$. Thus $E_{i*} = \Sigma\ e_{ik}E_{k*}$.
Let $G_{i*} = e_{ii}{}^{n}\ E_{i*}$, $G_{j*} = E_{i*}$ for $j \neq i$ define G,
where $e_{ii}{}^{n}$ is an idempotent power of e_{ii}. We will
show G is idempotent, and has the same row space as E.
For any j substitute the equation $E_{i*} = \Sigma\ e_{ik}E_{k*}$ into
$E_{j*} = \Sigma\ e_{jk}E_{k*}$. Since $e_{ji}e_{ik} \leq e_{jk}$ we have $E_{j*} =$
$\Sigma_{k\neq i}\ e_{jk}E_{k*} + e_{ji}e_{ii}E_{i*}$. Repeat the substitution n times.
Then $E_{j*} = \Sigma_{k\neq i}\ e_{jk}E_{k*} + e_{ji}e_{ii}{}^{(n)}E_{i*}$. This proves that
G spans the same row space (take $j = i$) and that for
$j \neq i$, $G_{j*} = g_{jk}G_{k*}$. We have $G_{i*} = e_{ii}{}^{(n)}E_{i*} =$
$\Sigma_{i\neq k}\ e_{ii}{}^{(n)}e_{ik}E_{k*} + e_{ii}{}^{(n)}E_{i*} = \Sigma_{i\neq k}\ g_{ik}G_{k*} + g_{ii}G_{i*}$.
Thus, G is idempotent. Continue this process until we
arrive at an idempotent E satisfying conditions (i),
(iii). Condition (iii) implies condition (iv) (consider the ith entry).

Now suppose that the rows of G do not form a stan-
dard basis. Let $G_{i*} = \Sigma\ c_{ij}G_{j*}$ for some c_{ij} where G_{i*}
$> c_{ii}G_{i*}$. By repeatedly substituting this expression
into itself we obtain a similar expression with
c_{ii} replaced by an idempotent power $c_{ii}{}^{n}$. Let H_{i*}
$= c_{ii}{}^{(n)}G_{i*}$ and $H_{j*} = G_{j*}$ for $j \neq i$. Then $R(H) = R(G)$
since G_{i*} can be expressed in terms of $c_{ii}{}^{n}\ G_{i*}$ and
other rows. Since $G_{i*} > c_{ii}{}^{n}\ G_{i*}$ and $G_{i*} = g_{ii}G_{i*}$

we must have $c_{ii}{}^n g_{ii} < g_{ii}$. We have $H_{i*} = c_{ii}{}^n G_{i*}$

$= c_{ii}{}^n \Sigma g_{ij} G_{j*} = \Sigma h_{ij} H_{j*}$. And for $j \neq i$, $H_{j*} =$

$G_{j*} = \Sigma g_{jk} G_{k*} \geq \Sigma h_{jk} H_{k*}$. But $G_{j*} = g_{jj} G_{j*}$. So eq-

uality holds. Continuation of these two processes

gives the desired idempotent. Since the idempotent is

decreased each time and 7 is finite, they must termi-

nate. $\square$

EXAMPLE 3.2.4.   Apply this process to the matrix

$$\begin{bmatrix} 0.4 & 0.6 \\ 0.3 & 1 \end{bmatrix}$$

then we obtain the matrix

$$\begin{bmatrix} 0.4 & 0.4 \\ 0.3 & 1 \end{bmatrix}$$

in which there is a standard basis.

THEOREM 3.2.4.   *If idempotents are linearly ordered*
then the row and column ranks are equal.

Proof.   The processes above do not change the row
or column rank, which are D-class invariants. Eventually we get a matrix whose row rank and column rank
each equal the number of nonzero main diagonal entries
$\square$

EXAMPLE 3.2.5.   For a 4-element lattice

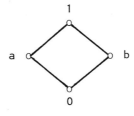

the matrix

$$\begin{bmatrix} a & a \\ b & b \end{bmatrix}$$

is idempotent. Its row rank does not equal its column rank.

DEFINITION 3.2.2.   A *cycle* in a Boolean matrix A is a sequence $i_1, i_2, \ldots, i_{r+1} = i_1$ such that $a_{i_j i_{j+1}} = 1$ for $j = 1$ to r. Its *length* is r.

THEOREM 3.2.5.   *Let E be an idempotent whose non-zero rows form a standard basis. Then* $e_{ii}E_{i*} = E_{i*}$, *main diagonal entries are idempotent, and each main diagonal entry is greater than or equal to any other entry in its row.*

Proof.   We have $E_{i*} = \Sigma\, e_{ik}E_{k*}$. By the standard basis property $E_{i*} = e_{ii}E_{i*}$. This implies $e_{ii}^{(2)} = e_{ii}$ and $e_{ij} = e_{ii}e_{ij} \le e_{ii}$. $\square$

PROPOSITION 3.2.6.   *If the idempotents in 7 are linearly ordered and F is the Boolean matrix such that* $f_{ij} = 1$ *if* $e_{ij} = e_{ii} > 0$ *and* $f_{ij} = 0$ *if* $e_{ij} < e_{ii}$, *and E is an idempotent matrix whose nonzero rows form a standard basis then*

$$\sum_{n=0}^{\infty} F^n$$

is antisymmetric.

Proof.   We will show that F has no cycles of length greater than 1. Suppose not. Let $i_1, i_2, \ldots, i_r$, $i_{r+1} = i_1$ be such that $e_{i_j i_{j+1}} = e_{i_j i_j}$ for $j = 1$ to $r + 1$. Then $E_{i_j *} \ge e_{i_j i_{j+1}}E_{i_{j+1}*} = e_{i_j i_j}E_{i_{j+1}*}$ for each j. By the above theorem, all $e_{i_j i_j}$ are idempotent.

Let $e_{i_j i_j}$ be the least of the $E_{i_j}*$. Then we have

$E_{i_j}* \geq e_{i_j i_j} E_{i_{j+1}}*$ and $E_{i_{j+1}}* \geq e_{i_{j+1} i_{j+1}} e_{i_{j+2} i_{j+2}}$ $\cdots$

$e_{i_{j-1} i_{j-1}} E_{i_j}*$. Multiply this equation by $e_{i_j i_j}$. The

we have $e_{i_j i_j} E_{i_{j+1}}* \geq e_{i_j i_j} E_{i_j}* = E_{i_j}*$. Thus row E_{i_j}

is dependent. This is a contradiction. $\Box$

3.3 TRIANGULAR FORM

In the fuzzy and Boolean cases it is important to put
idempotents into triangular form. The following exam
ple shows that it is not in general possible for matr
ces to be put into triangular form over any 7.

EXAMPLE 3.3.1. Let X generate a cyclic incline wi
$1 > x > x^2$. Then this is an idempotent.

$$\begin{bmatrix} 1 & x^2 & x \\ x & x & x^2 \\ x^2 & 1 & 1 \end{bmatrix}$$

THEOREM 3.3.1. *For any idempotent matrix $M \in M_n(7$*
over a linearly ordered incline the relation $\{(i, j):$
$m_{ij}^{n-1} > m_{ij}\}$ *is acyclic. Therefore, its transitive*
closure is a strict partial order.

Proof. Suppose not. Then for $k \leq n$, for some $i(1$
$i(2), \ldots, i(k)$ we have a cycle $m_{i(1)i(2)}^{n-1} >$

$m_{i(2)i(1)}, m_{i(2)i(3)}^{n-1} > m_{i(3)i(2)}, \ldots, m_{i(k)i(1)}^{n-1}$
$> m_{i(1)i(k)}$. By cyclic symmetry we may assume

$m_{i(k)i(1)} \leq m_{i(r)i(r+1)}$ for $r = 1, 2, \ldots, k - 1$.

Then $m_{i(1)i(k)} \geq m_{i(1)i(2)} m_{i(2)i(3)} \cdots m_{i(k-1)i(k)} \geq$

$i(k)i(1)^{n-1}$. This is a contradiction.   □

EXAMPLE 3.3.2.   This condition is necessary. Let P be an n-cycle and let $1 > x > \ldots > x^{n-1}$. Then $I + xP + \ldots + x^{n-1}P^{n-1}$ is idempotent and the relation $a_{ij}^{n-1} a_{ji}$ has an n-cycle (cycle of order n).

EXAMPLE 3.3.3.   The incline must be linearly ordered. Take a Boolean algebra of large order. Let main diagonal elements be 1. Then as long as $x_{ij} x_{rs} = 0$ for all pairs $\{i, j\} \neq \{r, s\}$ we can realize any antisymmetric irreflexive binary relation by $m_{ij}^{(n-1)} > m_{ji}$.

COROLLARY 3.3.2.   *Under the hypotheses of the theorem there exists a permutation matrix P such that E = MP^T has the property $x_{ij}^{(n-1)} \leq x_{ji}$ for $i < j$.*

THEOREM 3.3.3.   *Let S be order isomorphic to a product of a finite number of linearly ordered sets. Then for every idempotent matrix E there exists P such that $P^T = P^T P = I$ and $F = PEP^T$ satisfies $f_{ij}^{(n-1)} \leq f_{ji}$.*
    Proof.   Apply the previous result to get a permutation matrix in each factor of S. Putting them together gives P.   □

There is also a result on decomposability of matrices over a finite incline.

DEFINITION 3.3.1.   A matrix M in $M_n(7)$ is *g-decomposable* if there exists $\underline{s} \subset \underline{n}$ such that $m_{ij} < 1$ for $\varepsilon \underline{s}$, $j \varepsilon \underline{\tilde{s}}$, where $\underline{s}, \underline{\tilde{s}} \neq \emptyset$.
    A matrix G in $M_n(7)$ is called a *generalized inverse* (*-inverse*) of M, denoted by M^- if $M = MGM$.

EXAMPLE 3.3.4.   This matrix is not g-decomposable.

$$\begin{bmatrix} 1 & 0 & 1 \\ 1 & 0 & 0 \\ 0 & 1 & 0 \end{bmatrix}$$

THEOREM 3.3.4.   *Every idempotent over a Type I 7 except J is g-decomposable.*

Proof.   Let E be idempotent over 7. If E is not decomposable then the image of E in $M_n(\beta)$ under any epimorphism to $\{0, 1\}$ is indecomposable. Thus its image is J. Thus the image in any Boolean algebra is J. Thus the image in S is J. Here S is the same as in Proposition 2.2.2. Therefore E = J.   □

EXAMPLE 3.3.5.   For E

$$\begin{bmatrix} 1 & x & x \\ x & 1 & x^2 \\ x^2 & x & 1 \end{bmatrix}$$

a g-decomposition occurs where $\underline{s} = \{1\}$.

EXAMPLE 3.3.6.   Suppose we write an incline in addi tive notation so that the operations are inf $\{x, y\}$ an x + y (addition in a commutative semigroup). Then a matrix which is reflexive becomes one with zeros on it main diagonal, where 0 is the lowest element. Such a matrix is idempotent if and only if $f_{ij} + f_{jk} \geq f_{ik}$. This says that f is a directed metric, i.e., the direc ted distance in a graph. Such a metric can be obtaine from any matrix M by taking a power M^n over 7.

3.4 REGULARITY

In this section we are concerned with algorithms related to regularity.

For a Boolean matrices, the row space must be a distributive lattice. However, this is not true for general inclines.

EXAMPLE 3.4.1.   The matrix

$$\begin{bmatrix} 1 & 1 \\ 1 & 1 \end{bmatrix}$$

is idempotent and its row space is *7*, which may not be distributive lattice.

EXAMPLE 3.4.2.   This fuzzy matrix has regular sections but is not regular.

$$\begin{bmatrix} 1 & 0.4 & 0.3 \\ 0.5 & 1 & 0.4 \\ 0.5 & 0.5 & 1 \end{bmatrix}$$

Since the rows are a standard basis, if it were regular there would be a permutation which would be a g-inverse, but there is none.

In the case that idempotents in the incline are linearly ordered there exists the following algorithm to find a g-inverse of a matrix X.

ALGORITHM 3.4.1.   *(i) Find a standard basis for the row space. (ii) Let A be a matrix whose nonzero rows are the members of this row space. (iii) The matrix X is regular if and only if APA = A for some permutation matrix P. (iv) Choose a matrix T such that TX = A. (v) Then PT is a g-inverse of X.*

EXAMPLE 3.4.3. Consider the matrix over the incli [0, 1] under sup $\{x, y\}$ and xy.

$$\begin{bmatrix} x & x^2 \\ 1 & x \end{bmatrix}$$

(i) A standard basis is $(1 \quad x)$.

(ii) Let

$$\begin{bmatrix} 1 & x \\ 0 & 0 \end{bmatrix}$$

(iii) A permutation g-inverse is given by

$$\begin{bmatrix} 1 & 0 \\ 0 & 1 \end{bmatrix}$$

(iv)

$$\begin{bmatrix} 0 & 1 \\ 0 & 0 \end{bmatrix}\begin{bmatrix} x & x^2 \\ 1 & x \end{bmatrix} = \begin{bmatrix} 1 & x \\ 0 & 0 \end{bmatrix}$$

(v) This is a g-inverse of the given matrix:

$$\begin{bmatrix} 0 & 1 \\ 0 & 0 \end{bmatrix}$$

THEOREM 3.4.1. *If the idempotents are linearly ordered Algorithm 3.4.1 is valid.*

Proof. Let X be a regular matrix. Let Z be obtai ed by setting all rows other than those in a row basi equal to zero. Then there exists an idempotent E wit the same row space, whose nonzero rows contain a stan dard basis, by what we have shown.

Let A be as in the algorithm. Since a standard basis is unique $A = P^T E$ for some permutation P^T. So $APA = P^T EPP^T E = P^T EE = P^T E = A$. So P is a permutatio g-inverse of A.

Since R(A) = R(X), TX = A and SA = X for some T, S.
XPTX = SAPTSA = SAPA = SA = X.   ☐

EXAMPLE 3.4.4.   The following example shows that
unless idempotents are linearly ordered there may not
exist a g-inverse which is a partial permutation.   As
incline take the distributive lattice.

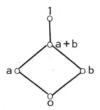

The matrix

$$\begin{bmatrix} a & b \\ b & a \end{bmatrix}$$

is regular since it is its own g-inverse.   But neither
of these is a g-inverse:

$$\begin{bmatrix} 1 & 0 \\ 0 & 1 \end{bmatrix}, \begin{bmatrix} 0 & 1 \\ 1 & 0 \end{bmatrix}$$

The permutation matrix P, if it exists, can be computed by the following algorithm.

ALGORITHM 3.4.2.   *Let A be a regular matrix in* $M_n(7)$,
of rank n. (i) *For a g-inverse to exist each row must
have a unique maximal element, which is idempotent, and
which divides all the other row elements.* (ii) *Form a
matrix F by* $f_{ij} = 1$ *if and only if* a_{ij} *equals the largest entry in* A_{i*}, *for a nonzero row* A_{i*}. (iii) *Let P
be the unique permutation less than or equal to F.
Then* P^T *is a g-inverse.*

EXAMPLE 3.4.5.   Let

$$M = \begin{bmatrix} 0.3 & 1 \\ 1 & 0.4 \end{bmatrix}$$

over the fuzzy algebra.

(i) This step is valid.

(ii)

$$F = \begin{bmatrix} 0 & 1 \\ 1 & 0 \end{bmatrix}$$

(iii)

$$P = \begin{bmatrix} 0 & 1 \\ 1 & 0 \end{bmatrix}$$

THEOREM 3.4.2.   *Algorithm 3.4.2 is valid if the idempotents are linearly ordered.*

Proof. We may as well rearrange the rows of A and assume we have an idempotent E. Step (i) has been proved earlier.

It is clear that I ≤ F and EIE = E. Moreover, if there were any other permutation P ≤ F then F would not be acyclic (i.e., would have a cycle). □

A matrix A is *regular* if and only if ABA = A for some matrix B. For basic concepts of regularity, see Clifford and Preston (1961). One way to test for regularity is to compute the greatest subinverse of a matrix, introduced in the Boolean case by Schein.

DEFINITION 3.4.1.   The *greatest subinverse* of a matrix A is the sum of all B such that ABA ≤ A.

EXAMPLE 3.4.6.   The greatest subinverse of

$$\begin{bmatrix} 1 & 0 & 1 \\ 1 & 1 & 0 \\ 0 & 1 & 1 \end{bmatrix}$$

is zero, if $1 = 1_3$.

The greatest subinverse exists and is unique for a compact incline by definition. Moreover, it can readily be computed.

PROPOSITION 3.4.3.   *A matrix M is regular if and only if MXM = M where X is the greatest subinverse of M.*

Proof.   Sufficiency is immediate. Let M be regular, X its greatest subinverse. Then MYM = M for some Y. But X ≥ Y so MXM ≥ M. Yet MXM ≤ M by definition of greatest subinverse, and the distributive law.   □

PROPOSITION 3.4.4.   *The (i, j)-entry x_{ij} of the greatest subinverse of M is the sum of all elements y of the incline such that for every k, r, $m_{ki}m_{jr}y \le m_{ir}$*

Proof.   Any matrix Y satisfies MYM ≤ M if and only if $m_{ki}y_{ij}m_{jt} \le m_{kt}$.   □

This gives a test for regularity of a matrix over any compact incline, and XMX will be the greatest *Vagner inverse* of M if one exists. A matrix G is said to be a *Vagner inverse* of A, denoted by A^* if A = AGA and G = GAG.

EXAMPLE 3.4.7.   Consider the matrix M which is

$$\begin{bmatrix} 1 & 0 & 0 \\ t & 1 & 0 \\ 0 & t & 1 \end{bmatrix}$$

over a cyclic incline generated by t with 0, 1.

Let X be the greatest subinverse. Then $IXI \leq MXM \leq M$. So X has zeros where M does. This takes care of the equations involving both diagonal entries on the left. Equations involving a zero entry on the left ar trivial. The remaining nontrivial equations involve a least one t, so will be satisfied unless $m_{ir} = 0$. Thi is automatic unless $i = 3$, $r = 1$. Thus $m_{3i} m_{j1} y_{ij} = 0$. Thus for $i = 2, 3$, $j = 1, 2$, $y_{ij} = 0$. So the greatest subinverse is

$$\begin{bmatrix} 1 & 0 & 0 \\ 0 & 0 & 0 \\ 0 & 0 & 1 \end{bmatrix}$$

Chen-Zhong Luo (1982) has provided an algorithm for finding all g-inverses of a fuzzy matrix. He does thi by finding a complete set of minimal g-inverses.

Over a general incline (nonlinear) symmetric idem-potents need not have a very special form. This is not true for the Boolean and fuzzy case.

EXAMPLE 3.4.8.   These matrices are idempotent where x, y, z generate an 8-element Boolean algebra and $1 > v > v^2$. Let $x + y + z = 1$.

$$\begin{bmatrix} 1 & v & v^2 \\ v & 1 & v \\ v^2 & v & 1 \end{bmatrix} , \begin{bmatrix} 1 & v & v^2 \\ v & 1 & v \\ v^2 & v^2 & 1 \end{bmatrix} , \begin{bmatrix} 1 & x & z \\ x & 1 & y \\ z & y & 1 \end{bmatrix}$$

3.5 MOORE-PENROSE INVERSES

Special types of g-inverses are sometimes used.

DEFINITION 3.5.1.   Let S be a semigroup. An *anti-automorphism* of S is a 1-1 onto function $\tau(a)$ such that $\tau(ab) = \tau(b)\tau(a)$.

EXAMPLE 3.5.1.   Transpose of a matrix gives an anti-automorphism of matrices over any commutative semiring.

DEFINITION 3.5.2.   An element x of a semigroup with antiautomorphism is *symmetric* if $x = \tau(x)$.

EXAMPLE 3.5.2.   A symmetric matrix is one case of this.

EXAMPLE 3.5.3.   If S is a group and $\tau(a) = a^{-1}$ then a symmetric element is one such that $a^2 = e$ where e is the identity element of S.

DEFINITION 3.5.3.   Let $x \in S$ where S is a semigroup with automorphism. Then y is a *least squares g-inverse* of x if and only if xyx = x, yxy = y, and $\tau(yx) = yx$ It is a *minimum norm g-inverse* if and only if xyx = x, yxy = y, and $\tau(xy) = xy$. It is a *Moore-Penrose inverse* if and only if both conditions hold.

If a matrix has a minimum norm (least squares) g-inverse then it lies in the L (R)-Green's equivalence class of a symmetric idempotent (which itself has a Moore-Penrose inverse). For matrices over R, Moore-Penrose inverses are important in giving least squares estimates of coefficients in a linear relation y = cx. In fact, yx^+ gives the estimate where x^+ is the Moore-Penrose inverse of x. Every matrix over R has a unique Moore-Penrose inverse. The other inverses also have statistical applications.

EXAMPLE 3.5.4. These matrices over $\mathbf{R}$ are Moore-Penrose inverses of one another.

$$\begin{bmatrix} 1 & 1 & 1 \end{bmatrix}, \qquad \begin{bmatrix} \frac{1}{3} \\ \frac{1}{3} \\ \frac{1}{3} \end{bmatrix}$$

THEOREM 3.5.1. *In any semigroup with antiautomorphism: (i) the Moore-Penrose inverse is unique if it exists; (ii) if it exists then the H-Green's equivalence classes of XX^T and X^TX are groups and $X^+ = (X^TX)^{-1}X^T = X^T(XX^T)^{-1}$; (iii) it exists if and only if $X^TX \, L \, X \, R \, XX^T$ if and only if $X \, H \, XX^TX$.*

    Proof. Suppose $X^TX \, L \, X \, R \, XX^T$. Then $X^TX \, R \, X^T \, L \, XX^T$ by transposing these relations. These imply $XX^TX \, R \, X$ and $XX^TX \, L \, X$. Suppose $XX^TX \, H \, X$. Then $X^TXX^T \, H \, X^T$. These imply $XX^TXX^T \, H \, XX^T$. This implies by Green's Lemma that the H-class of XX^T is a group. Define $X^+ = (X^TX)^{-1}X^T$. Then X^+X is an idempotent in the H-class of X^TX and by calculation XX^+ is also an idempotent.

    Suppose the Moore-Penrose inverse X^+ exists. Then $XX^+X = X$ and $X^+XX^+ = X^+$. Thus $X \, R \, XX^+ \, L \, X^+$ and $X \, L \, X^+X \, R \, X^+$. But XX^+ is symmetric. Thus $X^T \, L \, (XX^+)^T = XX^+$ and $X^T \, R \, (X^+X)^T = X^+X$. Thus $X^T \, L \, X^+$ and $X^T \, R \, X$. Therefore $X^TX \, L \, X^+X \, L \, X$ and $X \, R \, XX^T$. This proves necessity of (iii). We have $XX^T \, R \, XX^+$. But $XX^+ \, L \, X^+$ and because of (iii) $XX^T \, L \, X^T$. So $XX^T \, H \, XX^+$. Likewise $X^TX \, H \, X^+X$. So the H-classes of XX^T, X^TX are groups. Thus $(X^TX)(X^TX)^{-1} = X^+X$ and $(XX^T)(XX^T)^{-1} = XX^+$. The elements XX^+ and X^+X are unique since groups have unique idempotents. By Green's Lemma the mappings of multiplication by X is an isomorphism so

X^+ is unique. And in fact $(X^TX)^{-1}X^TX = X^+X$ so $(X^TX)^{-1}X^T = X^+$. The rest of (ii) can be verified similarly. $\square$

EXAMPLE 3.5.5. These are Moore-Penrose inverses:

$$\begin{bmatrix} 1 & 1 & 0 \\ -1 & 1 & 0 \\ 0 & 0 & 0 \end{bmatrix}, \quad \frac{1}{2}\begin{bmatrix} 1 & -1 & 0 \\ 1 & 1 & 0 \\ 0 & 0 & 0 \end{bmatrix}$$

THEOREM 3.5.2. *Any finitely generated group G of matrices over an incline is finite, and for all A, B ε G it is false that A > B.*

Proof. Consider the incline of all matrices whose entries lie in the incline generated by the generating matrices. It is a finitely generated incline containing all members of G. Every infinite sequence over it $a_1, a_2, \ldots$. has some $a_i \geq a_j$ for $i > j$. Let E be the idempotent in G. If $A > B$ in G however, $B^{-1}A > E$ so powers of $(B^{-1}A)$ give an infinite ascending chain. This is impossible. So every infinite sequence has only finitely many distinct elements. $\square$

THEOREM 3.5.3. *If X over an incline has a Moore-Penrose inverse X^+ then $X^+ = X^T$. If any matrix of the form $(XX^T)^n$, $n > 0$ is idempotent then XX^T is idempotent.*

Proof. Suppose $Y = X^+$ exists. We know it lies in the H-class of X^T. It would suffice to show that XX^T is idempotent since XX^T H XX^+ and multiplication by X is 1-1 on the H-class of X^+. We know the H-class of XX^T is a group. But since the H-class of XX^T is a group we know that $(XX^T)^n = X^TX$ for some $n > 0$, by the last theorem. Assume $n > 3d^2$ where d is the dimension of the matrices. We will show that for sufficiently

large m, $(XX^T)^m \leq (XX^T)^{m+1}$. For each term in the
expansion of $(XX^T)^m$ we can get a term in the expansion
of $(XX^T)^{m+1}$ by replacing x_{ij} by x_{ij}^3. The terms in
$(XX^T)^m$ for the (i, j)-entry have the form $\Pi \, x_{ij}^{n(i,j)}$
where $\Sigma \, n(i, j) = 2m$. Consider the terms as expression
in the incline which is the d^2-fold Cartesian product
of the incline with generators symbols a_{ij} and no relations. For any infinite chain in this incline some
member is greater than or equal to a later member.
Let A be the matrix (a_{ij}) over this incline. Then in
the chain $(AA^T)^n$, $(AA^T)^{2n}$, ... some member $(AA^T)^{kn} \geq$
$(AA^T)^{rn}$, $k < r$. This means each term in $(AA^T)^{rn}$ is a
multiple of a term in $(AA^T)^{kn}$. Some variable must
occur to a degree at least two higher in $(AA^T)^{rn}$ than
$(AA^T)^{kn}$. Then we obtain a term in $(AA^T)^{kn+1}$ greater
than or equal to this term if we replace x_{ij} by x_{ij}^3.
This implies $(AA^T)^{kn+1} \geq (AA)^{rn}$. Thus

$$(XX^T)^{kn+1} = (XX^T)^{rn} = (XX^T)^{kn}$$

Since XX^T lies in a group, it is idempotent.

For the second statement, we have shown that if a
power $(XX^T)^n$ lies in a group then it is idempotent
since $(XX^T)^{kn} = (XX^T)^{rn}$ for some k, r and $(XX^T)^{kn+1} \geq$
$(XX^T)^{rn}$. But $(XX^T)^{kn+1}$ also lies in the group as do
all powers above the nth. So $(XX^T)^{kn} = (XX^T)^{kn+1}$.   ☐

ALGORITHM 3.5.1.   *(i) If $MM^TM = M$ then M^T is the*
Moore-Penrose inverse of M. (ii) Otherwise M has no
Moore-Penrose inverse.

EXAMPLE 3.5.6.   Let M be the fuzzy matrix

$$\begin{bmatrix} 0.5 & 1 \\ 0.6 & 0.5 \end{bmatrix}$$

Then $MM^{T}M = M$ so M^{T} is its Moore-Penrose inverse.

3.6 NONCOMMUTATIVE INCLINES AND ANTIINCLINES

We show next that a number of the foregoing results hold even if multiplication is not required to be commutative. Such inclines occur as sets of ideals in a ring, as semigroups of words ordered on various ways, and as semidirect products.

We show for finite noncommutative incline **7** the mapping $x \rightarrow x^{n}$ where $x^{n} = x^{n+1}$ gives an isomorphism on every group in $M_n(\mathbf{7})$. The image of this mapping in **7** is a distributive lattice.

In many semigroups, $ab = e$ implies $ba = e$ where e is an identity. For these we prove an inverse theorem. Results on standard bases go through as before. We prove for finitely generated noncommutative inclines that every sequence contains a descending sequence.

DEFINITION 3.6.1.   A *noncommutative incline* (*antiincline*) is a set **7** on which two binary operations are defined, satisfying;

    (1) *(Associative)*: $(x + y) + z = x + (y + z)$
                               $(xy)z = x(yz)$
    (2) *(Commutative)*: $x + y = y + x$
    (3) *(Distributive)*: $x(y + z) = xy + xz$
                               $(x + y)z = xz + yz$
    (4) *(Idempotent)*: $x + x = x$
    (5) *(Incline)*: $x + xy = x$, $y + xy = y$
    (6) *(Antiincline)*: $x + xy = xy$, $y + xy = xy$

It is said to have 0, 1 if there exist elements 0, 1 such that for all x, $0 + x = x$ and $1x = x1 = x$.

EXAMPLE 3.6.1.   The set of noncommuting words in k generators ordered by length and ordered

lexicographically when of equal length is a noncommutative incline. Here the ordering is linear and x + y
is the greater of the two words.

PROPOSITION 3.6.1.   *The two-sided ideals in any
semiring with 0 (semigroup) form a noncommutative
incline under the operations $K + L$ $(K \cup L)$, and $<KL>$
(KL) where $<KL>$ denotes the additive subsemiring
spanned by the set KL.*

   Proof.   The incline property follows from $KL \subset K$
and $KL \subset L$ because both are two-sided ideals.   □

THEOREM 3.6.2.   *Suppose $x^n = x^{n+1}$ for all x in a
noncommutative incline 7. The mapping $x \to x^n$ is an
incline homomorphism with image the set of idempotents.
Any two idempotents commute in the image semiring.*

   Proof.   Since $(x + y)^n \geq x^n$ and $(x + y)^n \geq y^n$ we
have $(x + y)^n \geq x^n + y^n$. And each term of $(x + y)^{2n-1}$
has at least n x's or at least n y's. So is less than
or equal to either x^n or y^n. So $(x + y)^n \leq x^n + y^n$.

   The product in the set of idempotents is $(xy)^n$.
This is distributive since $(x(y + z))^n = (xy + xz)^n =$
$(xy)^n + (xz)^n$. It equals the usual product if that
product is idempotent. For associativity $(xyz)^n \geq$
$((xy)^n z)^n$. Yet $((xy)^n z)^n \geq ((xyz)^n (xyz))^n = (xyz)^n$.
So $((xy)^n z)^n = (xyz)^n$ which by symmetry is $(x(yz)^n)^n$.
It is commutative since yx is a subsequence of $(xy)^2$
so $(yx)^n \geq (xy)^{2n} = (xy)^n$. And $(xy)^n \geq (yx)^n$ by
symmetry. For homomorphism property, we need to show
that $(xy)^n = (x^n y^n)^n$. But $(xy)^n \geq (x^n y^n)^n$. And
$(xy)^n = ((xy)^{2n})^n = ((xy)^n (xy)^n)^n \leq (x^n y^n)^n$.   □

We can consider vector spaces acted on the right or on the left.

DEFINITION 3.6.2.   The set V_n of n-tuples from a semiring has the *left (right) vector space structure* $(x_1, \ldots, x_n) + (y_1, \ldots, y_n) = (x_1 + y_1, \ldots, x_n + y_n)$, $a(x_1, \ldots, x_n) = (ax_1, \ldots, ax_n)$, $(x_1, \ldots, x_n)a = (x_1a, \ldots, x_na)$.

EXAMPLE 3.6.2.   The semiring 7 is a vector space over itself of dimension 1.

DEFINITION 3.6.3.   A *subspace* of V_n is a subset closed under addition and multiplication by elements of the semiring.

EXAMPLE 3.6.3.   The set $\{(x, x)\}$ is a subspace of V_2.

DEFINITION 3.6.4.   The *left (right) span* of a set of vectors S in V_n is the set of all finite sums $\Sigma x_i v_i$ ($\Sigma v_i x_i$) where $x_i \in 7$, $v_i \in S$.

EXAMPLE 3.6.4.   The span of $\{(1, 0), (0, 1)\}$ is V_2.

DEFINITION 3.6.5.   A vector v is *dependent (independent)* of a set S if $v \in <S>$ ($v \notin <S>$). A *basis* for a subspace is an independent spanning set.

EXAMPLE 3.6.5.   Vectors with exactly one 1 and other entries 0 are independent and form a basis for V_n.

DEFINITION 3.6.6.   A *standard basis* is one in which if $v_i = \Sigma\, c_j v_j$ then $v_i = c_i v_i$.

PROPOSITION 3.6.3.   *Over any finite noncommutative incline, for any basis there exists a standard basis formed by multiples of the given basis.*

Proof. This is essentially the same as in the commutative case. $\square$

We next prove that if $ab = e$ then $ba = e$ provided that a semigroup is compact or satisfies the descending chain condition. This holds in finite semigroups and for matrices over a field.

THEOREM 3.6.4.   *Let S be a semigroup with identity e such that S is contained in a semigroup T with the same identity where T is compact or satisfies the descending chain condition on right ideals. Then if $ab = e$ in S, also $ba = e$.*

Proof. It suffices to prove the result where S satisfies one of the conditions (let $S = T$). Suppose first S satisfies the descending chain condition. Then $Sa^n = Sa^{n+1}$ for some n. So $a^n = xa^{n+1}$ for some x. Then $a^n b^n = xa^{n+1}b^n$ so $e = xa$. Now $b = xab = x$ so b is an inverse of a.

Next suppose S is compact. Let $S_1 = \{a^i : i > 1\}$. Let $\overline{S}_1$ be the closure of S_1. Then $\overline{S}_1$ is a compact semigroup. By Zorn's Lemma and compactness it has a minimal compact subsemigroup T_1. By minimality, for $x \in T_1$, $xT_1 = T_1 x = T_1$. Let $x \in T_1$. Choose f such that $xf = x$. Choose y such that $f = yx$. Then $f^2 = yxf = yx = f$. So f is idempotent. Let $f = \lim a^{n_i}$. Choose a subsequence (or subset) m_i of n_i such that b^{m_i} converges to some element b_1 and a^{c_i} converges to some element a_1 where $c_i = m_i - 1$. Then $a_1 a = a a_1 = f$. And $f b_1 = e$. Yet $f b_1 = f^2 b_1 = f e = f$. So $f = e$. So $a_1 a = a a_1 = e$. It follows that $a_1 = a_1 a b = b$. □

3   **EXAMPLE 3.6.6.**   The regular semigroup of transformations on $Z^+ F_\infty$ has a, b with $ab = e$, $ba \neq e$. For $a(x) = x + 1$, and $b(1) = 10$, $b(x) = x - 1$ for $x > 1$. Then $ab(x) = (x + 1) - 1 = x$. Yet $ba(1) = 11$.

This shows that regularity is not sufficient for $ab = e$ and $ba = e$ to be equivalent. Next we consider existence of inverses for matrices over a noncommutative semiring.

EXAMPLE 3.6.7.   We can let A be any $n^2 \times n^2$ partitioned permutation matrix. Let a_{ij} be an $n \times n$ matrix with precisely one 1 entry, in location $f(i, j)$, $g(i, j)$. Then the conditions are that f is 1-1 in j and g is 1-1 in i. For instance this matrix has an inverse

$$\begin{bmatrix} e_{12} & e_{21} \\ e_{11} & e_{22} \end{bmatrix} = \begin{bmatrix} 0 & 1 & 0 & 0 \\ 0 & 0 & 1 & 0 \\ 1 & 0 & 0 & 0 \\ 0 & 0 & 0 & 1 \end{bmatrix}$$

Here $b_{ij} = a_{ji}^*$ and e_{ij} is a matrix whose unique 1 entry is in location (i, j).

DEFINITION 3.6.7.   A semiring is *m-simple* if, as a module over itself, it is not a direct sum.

This is equivalent to saying there do not exist idempotents e_i with $\Sigma e_i = 1$ and $e_i e_j = 0$ for $i \neq j$.

EXAMPLE 3.6.8.   For $7 = [0, 1]$, $M_n(7)$ is m-simple.

THEOREM 3.6.5.   *Let A, B be matrices over a compact positive noncommutative semiring R.   If AB = I then BA = I.   Each b_{ij} is a Vagner inverse $a_{ji}{}^*$ of a_{ji}.*

If R is m-simple then there exists a permutation P such that b_{ji} and a_{ij} are 0 for $j \neq p(i)$ and $b_{ji} = a_{ij}{}^{-1}$.

Proof.   The first statement follows by the previous theorem.   Then we have $\Sigma\, a_{ij}b_{ji} = 1$, $a_{ij}b_{jk} = 0$, $i \neq k$ and $\Sigma\, b_{ij}a_{ji} = 1$, $b_{ij}a_{jk} = 0$, $i \neq k$.   So $a_{ik} = (\Sigma\, a_{ij}b_{ji})a_{ik} = 0 + a_{ik}b_{ki}a_{ki}$ and $b_{ik} = (\Sigma\, b_{ij}a_{ji})b_{ik} = b_{ik}a_{ki}b_{ik}$.   Therefore b_{ij} is $a_{ji}{}^*$.

The elements $e_j = a_{ij}b_{ji}$ satisfy $\Sigma\, e_k = 1$, $e_k e_j = 0$ for $k \neq j$.   So by m-simplicity some e_j is 1 and the others 0, for each i.   Thus there exists k such that $a_{ik}b_{ki} = 1$ and by the last theorem $b_{ki} = a_{ik}{}^{-1}$.   For $j \neq k$, $a_{ij}b_{ji} = 0$.   So $a_{ij}b_{ji}a_{ij} = a_{ij}$ is 0, and $b_{ji}a_{ij}b_{ji} = b_{ji} = 0$ for $j \neq k$.   If we work with $f_j = a_{ji}$ we find that the mapping $i \to k$ is a permutation. $\square$

LEMMA 3.6.6.   *In a poset the following are equivalent: (i) there exists no sequence w_n of distinct elements such that if $i < j$, $w_i \not\succ w_j$ and (ii) every infinite sequence of distinct elements has an infinite descending subsequence.*

Proof.   It is immediate that (ii) implies (i).   Suppose that (i) holds and let w_n be a sequence of distinct elements.   Then the set M of $\{n: w_n$ is maximal$\}$ is finite since if $i < j$ in it, $w_i \not\succ w_j$.   Every element is less than some maximal element.   Thus for some $w_k \in M$ there exists an infinite of elements w_i below it.   Thus look now at the sequence of $w_i < w_k$ such that $i > k$.   Repeat the process to find a second term $w_m < w_k$ and an infinite set of elements $w_i < w_m$.   This constructs an infinite descending subsequence and proves (ii). $\square$

We order the set of words in n generators by the relation that w > u if some elements of u can be deleted to give w. This is not an incline (addition is not defined) but if w ≥ u in this poset then whenever we substitute elements of any incline for the generators w ≥ u will hold. Moreover sums of these words will form an incline.

EXAMPLE 3.6.9.   Thus yy > xyxyx. Observe that in any incline y ≥ xyx and y ≥ yx.

LEMMA 3.6.7.   *Every infinite sequence of distinct words in n generators contains words W_i, W_j such that i < j and $W_i > W_j$.*

Proof.   We prove this by induction on the number of generators and for a fixed number of generators by the length of the first word W_1. For 1 generator $W_i > W_j$ whenever W_j is longer. For W_1 of length 1 we can reduce to the case of one less generator, since no other word can involve that generator.

Now consider W_1. Let its last letter be x_k.

Case 1. An infinite set of words W_i end in x_k. Then take this subsequence and delete x_k from each. We have reduced the problem to one in which W_1 has shorter length.

Case 2. An infinite set of words end in x_r, for some r ≠ k. Then look at the portion of the words behind the last x_k. By induction on the number of generators this has the property of the lemma. By the previous lemma there exists an infinite descending subsequence. Now in each case delete the last x_k and the portion beyond. By induction i < j and $W_i > W_j$

for some terms of the new sequence. But by construction this also holds in the original.   □

LEMMA 3.6.8.   *The poset $P(S)$ of lower ideals of a
poset S has no infinite nondescending chain if and
only if S has no system m_{ij}, $i < j$ such that $m_{ij} \nleq m_{jk}$.*
    Proof.   If W_i is a nondescending chain then W_j has
a monomial m_{ij} with $m_{ij} \nleq W_i$. So $m_{ij} \nleq m_{ki}$ for all k.
Conversely given m_{ij} let $W_j = \Sigma \ m_{ij}$. This gives a
nondescending chain.   □

LEMMA 3.6.9.   *If m_{ij} exist then they exist descending in j for fixed i.*
    Proof.   By taking subsequences for each i in turn
we can ensure that for m_{ij} is descending in i. This
uses the fact that the set of monomials has no nondescending chain.   □

DEFINITION 3.6.8.   Let u_i and v_i be two descending
sequences of monomials. Then they are *cofinal* if for
every i there exists j and k such that $u_i \geq v_j$ and
$v_i \geq u_k$.

EXAMPLE 3.6.10. $(xy)^n$ and $(yx)^n$ are cofinal since
$(xy)^n \geq (yx)^{n+1}$ and $(yx)^n \geq (xy)^{n+1}$.

THEOREM 3.6.10.   *For every descending sequence there
exists a cofinal sequence w_j of the form*
$$\prod_{i=1}^{k} z_i^{n_i(j)}$$
*where k is finite and for each i, $n_i(j)$ is non-
decreasing in j and z_i is a product of variables whose
subscripts are strictly increasing.*

Proof. We prove this by induction. For $n = 1$ it is true since either the sequence stops at x^c for some c or it is cofinal with x^i.

Consider n variables. If the number of x_1's which occur separated by all other variables but no other x_1 tends to infinity then the sequence is cofinal with $(x_1 x_2 \cdots x_n)^i$. In fact every word can be obtained from a high power of either of these by deletion.

Otherwise, we can write each word as $p_{i1} p_{i2} \cdots p_{ik}$ for some k where each p_{ik} involves not all variables. Suppose not. Let p_{ij} be obtained by starting by p_{ij-1} and stopping just before all variables occur. Then in $p_{ij-1} p_{ij}$ a sequence of all n distinct variables occurs. If arbitrarily many such sequences occur the sequence would be cofinal with $(x_1 x_2 \cdots x_n)^i$. In fact each $p_{ij-1} p_{ij}$ yields one variable of $(x_1 x_2 \cdots x_n)^i$. Take a subsequence such that each p_{ij} is descending in i. Now by induction the theorem follows. (Any descending sequence is cofinal with a subsequence, and being cofinal is transitive and products preserve cofinality). □

THEOREM 3.6.11. *In a finitely generated noncommutative incline every sequence has a descending subsequence.*

Proof. It suffices to prove this in the case of an incline of sums of monomials in n variables, since every n generator incline is a quotient of this. We assume the result is false. Assume that m_{ij} is descending. Let w_{ij} be a cofinal sequence having the form

$$\prod_{s=1}^{k} z_s^{n_s}$$

where z_s is a generator or a sequence of distinct generators in ascending order and n_s is constant or tends to infinity. We associate a single word W_i to each sequence W_{ij}. In all cases where the sequence n_s tends to infinity replace all powers $z_s^{n_s}$ by a new variable y_s depending only on z_s.

Let W_i be the resulting word, in the new variables as well as $x_1 x_2 \ldots x_n$. Then since the W_i are monomial in a finite set of variables, $W_i \geq W_j$ for some $i < j$. This implies that for all u there exists k such that $m_{iu} \geq m_{jk}$. Therefore $m_{ij} \geq m_{jk}$. This is a contradiction. $\square$

COROLLARY 3.6.12. *Every group in* $M_n(\mathbf{7})$ *consists of incomparable elements.*

Proof. If $A > B$ let $C = AB^{-1}$, $E = BB^{-1}$. Then $E < C < C^2 < \ldots$ has no descending subsequence. $\square$

COROLLARY 3.6.13. *Every group in* $M_n(\mathbf{7})$ *is finite.*

THEOREM 3.6.14. *For a matrix* $A \in M_n(\mathbf{7})$ *where* $\mathbf{7}$ *is a noncommutative incline, there exists* k *such that* $A^{k+d} \leq A^k$*, where* d *is any number divisible by* $1, 2,$ $\ldots, n$.

Proof. Choose m large enough that any collection of m numbers each at most n have a subcollection summing to d. It suffices that $\frac{m - n}{n}$ is at least d since then there will be enough of some one kind to do this. Let $k = nm$.

Now write any term of A^{k+d} as a product $a_{ii_1} \ldots a_{i_r j}$. Divide into disjoint segments of n factors. Each has a cycle. By deleting cycles we can reduce it to a term of length at most n. We can choose to delete exactly enough cycles to add to d by the condition on k. Therefore, we have a term of A^k which is greater than or equal to the given term of A^{k+d}. $\square$

COROLLARY 3.6.15.   *If A lies in a group in $M_n(\mathcal{I})$ then A has order dividing d.*

Proof.   Let $A^{k+d} \leq A^k$. If $A^{k+d} < A^k$ we have a contradiction to the last corollary. So $A^{k+d} = A^k$.   ☐

THEOREM 3.6.16.   *Over a noncommutative incline, for any m if A is idempotent then $a_{ij} = \sum_r a_{ir} a_{rr}{}^m a_{rj}$.*

Proof.   It is immediate that $a_{ij} = a_{ij}{}^{(m+2)} \geq \sum a_{ir} a_{rr}{}^m a_{rj}$. Consider any term in $a_{ij}{}^{(nm+2)} = a_{ij}$. Some index r must occur at least m times. The portion of the product between any two occurrences is at most $a_{rr}{}^{(k)} = a_{rr}$ for some k. The initial and final segments are at most $a_{ir}{}^{(s)}$ and $a_{rj}{}^{(t)}$. So this term is less than or equal to $a_{ir} a_{rr}{}^m a_{rj}$. So the sum of all terms is less than or equal to $\sum a_{ir} a_{rr}{}^m a_{rj}$. So equality holds.   ☐

THEOREM 3.6.17.   *The mapping $h: x \to x^n$ where $x^n = x^{n+1}$ in a noncommutative incline $\mathcal{I}$ induces a 1-1 mapping on each group in $\mathcal{I}$.*

Proof.   Let $X^n = E$ but $X \neq E$ where E is idempotent. It will suffice to show that $X \varepsilon$ ker h means $X = E$. We have $e_{ij} = \sum e_{ir} e_{rr}{}^m e_{rj}$ for $m \geq n$. So since $h(E) = h(X)$, $e_{ij} = e_{ir} x_{rr}{}^m e_{rj}$. Therefore $E \leq EX^m E$. Take m such that $X^m = X$. Then $E \leq EXE = X$. Since elements of a group are incomparable, $E = X$.   ☐

COROLLARY 3.6.18.   *Every finitely generated group in $M_n(\mathcal{I})$ lies in a product of symmetric groups S_n.*

Noncommutative continuous inclines, by Fuchs (1963), must have dimension greater than 1.

THEOREM 3.6.19. *A continuous incline structure on any finite interval* [a, b] *is commutative provided* $b^2 = b$.

Any semigroup is contained in the multiplicative semigroup of an incline.

DEFINITION 3.6.9.   Let S be a semigroup. The *semigroup semiring* T[S] for a semiring T is the set of all functions f: S → T with addition of functions and products given by fg(s) = $\sum_{uv}$ f(u)g(v).

EXAMPLE 3.6.11.   If Y is a ring and S a group this is the group ring.

DEFINITION 3.6.10.   The *semigroup incline* 7(S) for an incline 7 is the quotient of 7[S] by the congruence relation that x + xy = x, y + xy = y.

EXAMPLE 3.6.12.   Let 7 = β = {0, 1} is the incline consisting of all two-sided ideals of S.

DEFINITION 3.6.11.   A semiring S_1 is called a *semiring of endomorphisms* of S_2 if for each element a ε S_1 there exists a function f_a: S_2 → S_2 such that (i) $f_a(x + y) = f_a(x) + f_a(y)$, (ii) $f_a(xy) = f_a(x)f_a(y)$, (iii) $f_{a+b}(x) = f_a(x) + f_b(x)$, (iv) $f_{ab}(x) = f_b(f_a(x))$.

DEFINITION 3.6.12.   Suppose S_1 is a semiring of endomorphisms of S_2. Then the *semidirect product* of S_1 and S_2 is $S_1 \times S_2$ with operations (x, y) + (u, v) = (x + u, y + v) and (x, y)(u, v) = (xu, (u * y)v).

EXAMPLE 3.6.13.   If $y * u = u$ for all y we would have the ordinary direct product.

PROPOSITION 3.6.20.   The semidirect product of any two semirings is a semiring.

Proof. The most complicated property to prove is associativity: $((a, d)(b, e))(c, f) = (ab, (b * d)e)$ $\cdot (c, f) = (abc, (c * (b * d)e)f)$ and $(a, d)((b, e)$ $\cdot (c, f) = (a, d)(bc, (c * e)f) = (abc, ((bc) * d)$ $\cdot (c * e)f)$. But $(c * (b * d)e)f = (c * (b * d))(c * e)f = ((bc) * d)(c * e)f$. □

THEOREM 3.6.21.   There are the following semirings of endomorphisms: (i) 7_3 on 7_3 by $x^{1/a}$, (ii) 7_3 on 7_2 by $\frac{x}{a}$, (iii) 7_3 on 7_1 by $x^{1/a}$, (iv) 7_2 on 7_1 by $x - a$, (v) 7_1 on 7_1 by $\inf\{x, a\}$.

Proof. A computation. □

PROPOSITION 3.6.22.   The negative elements of any semilattice ordered group form a noncommutative incline.

Proof.   We need to show $x(y \vee z) = xy \vee xz$ and the incline property. Since $x(y \vee z) \geq xy$ and $x(y \vee z) \geq xz$, $x(y \vee z) \geq xy \vee xz$. Also $x^{-1}(xy \vee xz) \geq x^{-1}xy \vee x^{-1}xz = y \vee z$. Multiply by x. $xy \vee xz \geq x(y \vee z)$.

The incline property $xy \leq x$ and $yx \leq x$ follows from negativity. □

The positive elements under $\wedge$ form an incline also. For what semigroups is this valid ? Cancellation semigroups ?

EXAMPLE 3.6.14.   A nondistributive lattice is a semilattice ordered inverse semigroup but is not an incline.

In antiinclines (commutative or noncommutative inclines) we have xy ≥ x, xy ≥ y instead of xy ≤ x, xy ≤ y. For linear orders, there is a duality between inclines and antiinclines obtained by keeping the same multiplication but reversing the order. This also works for products of linear orders.

EXAMPLE 3.6.15. The dual to Boolean algebra β is an antiincline which we can write $\{1, \infty\}$ where
$$1 \cdot 1 = 1 + 1 = 1, \quad 1 + \infty = 1 \cdot \infty = \infty + \infty = \infty \cdot \infty = \infty.$$

DEFINITION 3.6.13. The *antiincline* $\{1, \infty\}$ is denoted 7^{-1}. The dual of an incline 7 (or more generally any antiincline) is denoted $7'$.

The semigroup $M_n(7^{-1})$ is simpler than $M_n(\beta) = B_n$.

LEMMA 3.6.23. In $M_n(7^{-1})$, $AB = XY$ where X is the vector of row sums of A and Y is the vector of column sums of B.

Proof. We have $(AB)_{ij} = 1$ if and only if for all k, $a_{ik} = b_{kj} = 1$ if and only if $\Sigma\, a_{ik} = 1$ and $\Sigma\, b_{kj} = 1$ if and only if $x_i = y_j = 1$. □

THEOREM 3.6.24. In $M_n(7^{-1})$ a matrix can be factored if and only if it is of rank 1. The only idempotents are J, ∞J. There are no nontrivial groups. If $B \neq J$, $ABC = \infty J$. The index of a matrix is at most 3.

Proof. Straightforward. □

Reflexive Boolean matrices (or matrices over 7) form an antiincline since $XY \geq XY + XY \geq XI + IY \geq X + Y$. In the next theorem let $7'$ be finitely generated.

THEOREM 3.6.25.   *In the antiincline $7'$ there cannot exist any infinite nonascending sequence, i.e., for any sequence w_i there exists $w_i < w_j$ with $i < j$.*

Proof.   Suppose not. It suffices to consider $7'$ with no relations. Then in $7'$ there exists a monomial m in w_i not less than or equal to any in w_j. In w_1 choose some monomial which occurs for an infinite collection of w_j and restrict to this subset. Do the same for the new w_2. Thus we obtain a sequence of monomials m_1, m_2, ... such that for $i < j$ it is false that $m_i \leq m_j$. Then Lemma 3.6.7 gives a contradiction. □

This proof applies also to noncommutative antiinclines.

Each of the inclines 7_1, 7_2, 7_3 has a dual antiincline.

EXAMPLE 3.6.16.   The antiincline $7_3' = (1, \infty)$ is the dual of 7_3 and its operations are sup, ×.

EXAMPLE 3.6.17.   The antiincline $7_1' = [1, \infty)$ is the dual of the fuzzy algebra and its operations are sup, sup.

EXAMPLE 3.6.18.   There is an incline for generators x_1, ... , x_k of all formal sums ΣM_i of noncommutative monomials in the x_i, where $M = M + N$ if N can be obtained by deleting variables from M.

DEFINITION 3.6.14.   A Boolean matrix M is *row nonempty* if and only if each row has at least one 1.

EXAMPLE 3.6.19.   This is row nonempty

$$\begin{bmatrix} 1 & 1 & 1 \\ 1 & 0 & 0 \\ 1 & 0 & 0 \end{bmatrix}$$

PROPOSITION 3.6.26.   *For any ring (semigroup) with unit the family of subspaces (subsets) containing 1 is an antiincline under $X + Y$, XY ($X \cup Y$, XY).*
The proof is a computation.

We omit all further proofs on antiinclines.

THEOREM 3.6.27.   *Over an antiincline bases are unique.*

There exists a mapping like a homomorphism from row nonempty matrices in $M_n(\beta)$ into $M_n(7')$ whenever $7'$ contains an element x where $1 < x < x^2$.

THEOREM 3.6.28.   *Let $f: \beta \rightarrow 7'$ be defined by $f(0) = 1$, $f(1) = x$ where $x > 1$. Then for row nonempty matrices A, B over β we have $f(A)f(B) = xf(AB)$.*

Matrices over an antiincline (unlike an incline) form themselves an antiincline.

THEOREM 3.6.29.   *The set $M_n(7')$ is a noncommutative antiincline*

COROLLARY 3.6.30.   *The semigroup $M_n(7')$ has no nontrivial groups.*

THEOREM 3.6.31.   *Over $7_1'$, $7_2'$, $7_3'$ for any matrix A there exist k_0, d, C such that $A^{k+d} = C \odot A^k$ for $k \geq k_0$.*

DEFINITION 3.6.15.   The *variety* generated by a class C of structures is the class of all structures obtained from C by taking (i) substructures, (ii) quotient structures, and (iii) direct products.

EXAMPLE 3.6.20.   The variety of rings generated by Z is the variety of all commutative rings.

THEOREM 3.6.32.   *The varieties generated by 7_2, 7_3 are equal and have defining relations*

$$M \leq \sum_{j=1}^{n+1} M_j$$

whenever the exponents of M as a vector lie in the convex hull of all vectors greater than or equal to the exponents of some M_j.

Proof.   We have epimorphisms $7_3 \to 7_2$ given by $\dfrac{\log x}{n}$. This gives a monomorphism $7_3 \to \overset{\infty}{\underset{1}{\times}} 7_2$ taking x to $\left(-\log x, -\dfrac{\log x}{2}, \ldots\right)$ so 7_2, 7_3 generate the same variety. As above, we need only consider log M $\leq \sup \{\log M_i\}$ or $\sum b_j y_j \leq \sup (\sum a_{ij} y_j)$ over all negative n-tuples or $\sum b_j y_j \geq \inf (\sum a_{ij} y_j)$ over all positive n-tuples. This is equivalent to $\sum b_j y_j \geq \inf \{z_j y_j\}$, $n \geq 1$ where the inf is over $z \geq A_{i*}$ for some i over all n-tuples y_j since if some y_j is negative the new set can be taken arbitrarily small. But this yields that (b_j) is a convex combination of vectors $z \geq A_{i*}$. $\square$

THEOREM 3.6.33.   *The variety of noncommutative semilattice ordered semirings generated by all inclines and antiinclines can be defined by the relations $M_1 \leq M_2 + M_3$ where M_1 can be obtained from M_3 and M_2 from M_1 by deleting certain variables.*

COROLLARY 3.6.34.   *The unique way to define a semilattice ordered semiring structure on $\{0, 1, \infty\}$ containing β, 7^{-1} and in the variety generated by incline and antiinclines is $0 \cdot \infty = \infty$.*

Proof.   Since $xy \quad x + xyz$, for $y = \infty$, $z = 0$, $x = 1$ this yields $\infty \leq 1 + 0 \cdot \infty$, or $0 \cdot \infty = \infty$. Conversely, the law $xy \leq x + xyz$ holds if x, y, or z is ∞ and otherwise it follows since it holds in β. $\square$

THEOREM 3.6.35.   *Every system of polynomial equations $\sum b_{ij} \prod x_i^{n_{ij}} = \sum a_{ij} \prod x_i^{m_{ij}}$ over 7_1, 7_2, 7_3 or $7_1'$, $7_2'$, $7_3'$ is finitely decidable.*

Proof.   The monomials on each side of every equation are linearly ordered. By dividing into cases, assume a specific linear ordering. This reduces the system to a system of inequalities and equalities on monomials. But this is a system of linear equations and inequalities on real number variables. It is known that these are decidable. $\square$

3.7 OPEN PROBLEMS

1. Give an algorithm to find all g-inverses of a matrix over an incline.

2. Give an algorithm to find all Vagner inverses of a matrix.

3. Does the row space of an idempotent matrix have an incline structure ?

4. Generalize Algorithm 3.4.2 to the case when zero 3 rows occur.

5. Study least squares g-inverses and minimum norm g-inverses over an incline.

6. Is it true that lattice ordered cancellation semigroups give inclines ?

Chapter 4

Topological Inclines

The order structure in many cases determines a compact Hausdorff topology on an incline. We characterize inclines order equivalent to [0, 1] and in some cases to $[0, 1]^n$. We study topological semilattices in general, showing they are usually homotopically trivial but may be nonmodular or nondistributive. We consider concepts of measure, integration, and Hilbert space for inclines.

4.1 LIMITS AND TOPOLOGY

For topological algebraic structures, the underlying set must have a topology, and the operations must be continuous. The latter condition can also be stated in terms of open sets.

All topologies in this book will be understood to be Hausdorff unless otherwise mentioned.

DEFINITION 4.1.1. A topology is *Hausdorff* if and only if for any two points x, y there exist disjoint open sets U, V such that x ε U, y ε V.

EXAMPLE 4.1.1.   Any topology defined by a metric (distance function) is Hausdorff.

Compact Hausdorff spaces and metric spaces are normal, and points are closed.

DEFINITION 4.1.2.   A topological space is *normal* if and only if for any two disjoint closed sets C_1, C_2 there exist disjoint open sets U_1, U_2 such that $C_1 \subset U_1$ and $C_2 \subset U_2$.
This implies that if U is open and x ε U there exists an open set V such that $\overline{V} \subset U$ and x ε V.

DEFINITION 4.1.3.   A topological space is *first-countable* if there exists a countable basis at every point.

EXAMPLE 4.1.2.   All metric spaces are first countable.

In a first countable Hausdorff space, at every point x there exists a sequence $V_1 \supset V_2 \supset \ldots \supset V_n$ of open sets whose intersection is $\{x\}$: let V_i be the intersection of the first i elements of a basis. We will need this property in order to work with countable sequences.

PROPOSITION 4.1.1.   *In any compact semilattice, every monotone sequence has a limit which equals its supremum or infimum.*

THEOREM 4.1.2.   *A first countable topology on a compact semilattice with 0 is unique.*

In the study of topological groups, a special measure called *Haar measure* plays an important role. We can define an analogous concept for topological inclines although we do not know conditions under which it exists or is unique.

A σ-*algebra* is a family of subsets of a set closed under complement and countable intersection. Let 7_A denote a σ-algebra of subsets of 7.

DEFINITION 4.1.4.   A *measure* on 7 is a function $m: 7_A \to R$ such that (i) $m(A) \geq 0$, (ii) $m(\overset{\infty}{\underset{1}{\cup}} A_i) = \overset{\infty}{\underset{1}{\Sigma}} m(A_i)$ if the A_i are pairwise disjoint, (iii) for some function $f(x) = m(xA) = f(x)m(A)$ for all sets A on which multiplication by x is 1-1.

EXAMPLE 4.1.3.   For 7_2 Lebesgue measure is a measure where $7_2 = [0, 1]$, inf $\{x, y\}$, and inf $\{x + y, 1\}$.

4.2 CHARACTERIZATION OF INCLINES ON [0, 1]

We start with the following simple proposition.

PROPOSITION 4.2.1.   *Suppose that in a continuous incline structure on* [0, 1] *a and b are idempotents. Then for all x between a and b, ax = a and bx = b.*

Proof.   We have $ax \geq a^2 = a$.   And $ax \leq a$.   So $ax = a$.   Let $f(x) = bx$.   Then $f(b) = b$, $f(a) = a$.   So for some y, $f(y) = x$.   Then $bx = b(by) = by = x$. $\square$

In the following we shall show how to classify all continuous incline structures on [0, 1] such that $0^2 = 0$, $1^2 = 1$.

DEFINITION 4.2.1.   The *index* of an element x ε 7 is the least k such that $x^k = x^{k+1}$.

EXAMPLE 4.2.1. An element has index 1 if and only if it is idempotent.

More generally, an element has finite index if and only if some power is idempotent (if $x < 1$ means some power is 0).

THEOREM 4.2.2. *Suppose in an incline structure on* $[0, 1]$ *both* $0, 1$ *are idempotent and that every other element has finite index greater than 1. Then the incline is isomorphic to* 7_2.

Proof. This follows from Theorem 9.2 of Fuchs (1963). □

THEOREM 4.2.3. *An incline structure on* $[0, 1]$ *in which* $0, 1$ *are the only idempotents and no other elements have finite index is isomorphic to* 7_3.

Proof. This also follows from Theorem 9.2 of Fuchs (1963). □

These results give a precise picture of all incline structures on $[0, 1]$ in which 1 is an identity. There exists a closed set of idempotents on which the incline is given by infimum and supremum. The complement consists of open intervals bounded by two idempotents. Therefore, we must have either the structure ($[0, 1]$, inf, +) or ($[0, 1]$, sup, ×) on each of them.

4.3 N-DIMENSIONAL LATTICES

In this section we study compact lattice structure on subsets of n-dimensional space. Several concepts from topology are needed.

DEFINITION 4.3.1. A *finite simplicial complex* is a compact metric topological space which is the union of sets homeomorphic to the n-dimensional tetrahedron $\{(x_1, \ldots, x_n): x_i \geq 0, \Sigma\, x_i \leq 1\}$ called *simplices* such that (i) the intersection of any two disjoint simplices is empty or again a simplex of lower dimension, and (ii) each face of a simplex is again a simplex.

EXAMPLE 4.3.1.   The n-cube $[0, 1]^n$ is a simplicial complex.

Any topological space which in some sense is composed of a finite number of pieces is a simplicial complex.

EXAMPLE 4.3.2.   A sphere is homeomorphic to an octahedron. This is a simplicial complex since it is made up of triangles any two of which intersect in $\emptyset$, a point or an edge.

DEFINITION 4.3.2.   A *tree* is a connected graph having no cycles.

EXAMPLE 4.3.3.   This is a tree

DEFINITION 4.3.3.   A *homotopy* between two functions f, g: X → Y is a continuous function h(t, x) from [0, 1] × X to Y such that h(0, x) = g(x) and h(1, x) = f(x).

EXAMPLE 4.3.4.   Any two functions from any space x to a convex set in R^n are homotopic by h(t, x) = tf(x) + (1 - t)g(x).

Homotopy is an equivalence relation.

EXAMPLE 4.3.5.   Let y be finite. Two mappings f, g are homotopic if and only if they are equal.

DEFINITION 4.3.4.   A topological space X is *contrac-tible* if and only if there exists a homotopy between the identity mapping x → x and a constant mapping x to x.

EXAMPLE 4.3.6.   A finite graph is contractible if and only if it is a tree.

EXAMPLE 4.3.7.   An n-cube is contractible, but an n-sphere (its boundary) is not.

A contractible space can be thought of as one which has no holes or loops or handles.

PROPOSITION 4.3.1.   *Let 7 be an compact Hausdorff incline such that the product mapping 7 × 7 → 7 is onto. Then 7 has 0, 1.*

Proof.   There exists a highest element 1 and a lowest element 0 by compactness. We have 1 · 1 = 1 since products are onto, so xy = 1 for some xy but then x, y must be 1. We have 0 · x ≤ 0 so 0 · x = 0 for all x. And 0 + x = x.

Let S = {x: 1 · x = x}. If x = 1 ·y then 1 · x = 1 · 1 · y = 1 · y = x. And if x ∈ S then x = 1 · x. So S = {1 · y}.

Suppose u ∉ S. Let x be a maximal element such that

$x \cdot y = u$ for some y.   By compactness such an x exists.
If $x = 1$ then $u \in S$.   If $x < 1$ then $x = rs$ for some
$r, s \geq x$.   So $r, s = x$ else $u = r(sy)$.   So $x = x \cdot x$.
So $1 \cdot x \geq x \cdot x \geq x$.   So $1 \cdot x = x$.   So $1 \cdot u = u$.
So $u \in S$.   This proves 1 is an identity element.   □

THEOREM 4.3.2. . *If G is continuous path connected
semilattice, all homotopy groups of G are zero, i.e.,
all mappings from an n-sphere S^n to G are homotopic.*

Proof.   The homotopy classes of mappings S^n to G
are known to form a group $\pi_n(G) = H$.   The mapping
$G \times G \to G$ gives an operation $H \times H \to H$ which is commutative, associative, and idempotent.   Moreover, it is
a group homomorphism.

Let H be any group with such a mapping f from $H \times H$
to H.   Then $f(x, y) = f(x, e)f(e, y) = g(x)g(y)$ where
g is the homomorphism $f(x, e)$.   Associativity means
$g(g(x)g(y))g(z) = g(x)g(g(y)g(z))$ or $g(g(x)) = g(x)$.
Since $g^2(x) = f(x, x) = x$, and $g(x)$ is onto.   So $g(x)$
is the identity.   So $g^2(x) = x$ means $x^2 = x$ in H.   So
H is trivial.   □

EXAMPLE 4.3.8.   Every tree is a semilattice, and
is contractible.

THEOREM 4.3.3.   *If a finite graph is a lattice, it
must be homeomorphic to $[0, 1]$.*

Proof.   By the previous theorem the graph must be
contractible.   This implies it is a tree.   For all x
we have disjoint paths x to zero and x to 1 given by
$x \vee h(t)$ and $x \wedge h(t)$ where $h(t)$ is a path from 0, 1.
This means the tree cannot have any endpoints except
0, 1.   □

Retracts are of importance in topology.

DEFINITION 4.3.5.   An subincline 7_1 of 7 is a *re-tract* if and only if there exists an incline homomorphism $7 \to 7_1$ which is the identity on 7_1.

EXAMPLE 4.3.9.   The subincline $S = \{x \colon x^2 = x\}$ is a retract of 7 by the mapping $x \to x^n$ of 7 is finite.

A continuous retract of a contractible space is contractible. Conversely, if A is a contractible sub-complex of a finite simplicial complex X, then A is a retract of X.

DEFINITION 4.3.6.   A *convex cone* in $\mathbf{R}^n$ is a set C such that for all $r \in \mathbf{R}^+$, $x, y \in C$ we have $x + y \in C$, $rx \in C$.

EXAMPLE 4.3.10.   The set of vectors X satisfying a finite number of inequalities $X \cdot a_{<i>} \geq 0$ for vectors $a_{<i>}$ is a convex cone.

DEFINITION 4.3.7.   A *continuous Hasse diagram* con-sists of (i) a subset C_p of $\mathbf{R}^n$, (ii) a fixed vector v, (iii) a choice of a closed convex cone f(x) at each point x of C_p such that for all $w \in f(x)$, $w \cdot v > 0$, and f(x) contains an open neighborhood of v, and (iv) the function f is upper semicontinuous.

EXAMPLE 4.3.11.   On $\mathbf{R}^n$ we can take v as any fixed vector and f(x) to be the constant function to a cone of vectors w such that w makes an angle of at most θ with v.

DEFINITION 4.3.8.   Let P be the set of x, y such
that there exists a continuous, piecewise differentiable
curve u(s) for s ε [a, b] for some a, b ε **R** such that
|u'(s)| = 1 and u(a) = x and u(b) = y and u'(s) lies
in the cone f(u(s)) for all s ε [0, 1]. Then P is the
*partial order associated with the continuous Hasse
diagram.*

PROPOSITION 4.3.4.   *Let P be the same as in the
above definition.   Then P is a partial order.*
   Proof.   Since we can take a = b, P is reflexive.
P is transitive because we can join a curve from y to
z to the end of a curve from x to y.   It remains to
show P is antisymmetric.   But this is true since if
(x, y) ε P and x $\neq$ y then (y - x) · v > 0 since
u'(s) · v > 0 for all s.   □

The partial order P is not in general closed.

EXAMPLE 4.3.12.   We can take the set C_p to be the
union of a collection of curves f_n from x to y_n such
that the limit of the functions f_n is not continuous.

DEFINITION 4.3.9.   A continuous Hasse diagram is
simplicial if P is a finite simplicial complex and f
is a function which is constant on the interior of each
simplex.

EXAMPLE 4.3.13.   For any simplicial complex contained
in R^n, if f is a constant we have a simplicial Hasse
diagram.

PROPOSITION 4.3.5.   *For a simplicial Hasse diagram
we may take the curves u(s) to be curves whose inter-*

section with any simplex is a straight line segment.

PROOF. This follows from the fact that if a curve exists from x to y within a simplex, then y lies in the cone at x, so the straight line segment from x to y is in the cone. □

THEOREM 4.3.6. *The partial order arising from a simplicial Hasse diagram is closed, as a subset of P × P.*

Proof. This is because a limit of simplicial paths in the required cones is again a simplicial path. □

The question, when is a simplicial Hasse diagram a lattice, is more subtle. We have seen that in this case the simplicial complex must be contractible.

A continuous lattice on [0, 1] × [0, 1] need not be distributive, or even modular.

EXAMPLE 4.3.14. The following is a simplicial partial order on a set homeomorphic to [0, 1] × [0, 1] which contains a pentagon. Take as underlying set a regular pentagon with one side vertical, and v as (0, 1). Subdivide the pentagon as pictured:

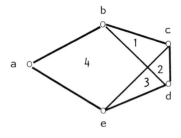

In regions 2, 3 let a side of the cone be parallel to de. Elsewhere let this side be parallel to ab. Let the other side of the cone be parallel to bc in

regions 1, 2 and to ae elsewhere. Then we have a continuous lattice such that a ∨ d = b, a ∧ c = e. Thus
a, b, c, d, e form a pentagon.

EXAMPLE 4.3.15.   It is also possible for a simplicial lattice to contain a diamond. Take a set in R^3
which at height t consists of the four line segments
from $(0, 0)$ to $(\pm 1 - |t|, \pm 1 - |t|)$ for $t \in [-1, 1]$.
Let $v = (0, 0, 1)$. Take as the constant cone $z^2 = x^2$
$+ y^2$. Then any two of the points $(\pm 1, \pm 1, 0)$ have
inf $(0, 0, -1)$ and sup $(0, 0, 1)$.

It is also possible to have a connected distributive
lattice which is not homeomorphic to $(0, 1]^n$.

EXAMPLE 4.3.16.   Let v be $[0, 1)$ and the cones have
sides at a $45°$ angle.

Then this is a distributive lattice, and therefore an
incline not homeomorphic to $[0, 1] \times [0, 1]$.

4.4 ONE PARAMETER SEMIGROUPS

The semigroups are related to continuous dynamical
systems and have other uses. For instance, they can
help in finding nth roots.

In this section we first take up square roots. We
generalize the condition det $(A) \geq 0$ if $\sqrt{A}$ exists to
arbitrary semirings in which $2xy \leq x^2 + y^2$.

DEFINITION 4.4.1.   A *1-parameter semigroup* is a continuous homomorphism $A(t)$ from $R^+ \cup \{0\}$ under addition to $M_n(\mathcal{T})$.

EXAMPLE 4.4.1.   This is a 1-parameter semigroup:

$$\begin{bmatrix} 0.2^t & 0 \\ 0 & 0.5^t \end{bmatrix}$$

DEFINITION 4.4.2.   For a matrix A, let

$$d_+(A) = \sum_{\pi \text{ even}} \prod_i a_{i\pi(i)}, \quad d_-(A) = \sum_{\pi \text{ odd}} \prod_i a_{i\pi(i)}$$

Therefore, per $(A) = d_+(A) + d_-(A)$. Here per (A) denotes the permanent of A.

THEOREM 4.4.1.   *Over any semiring $\mathcal{T}$ in which $2xy \geq x^2 + y^2$ for any $A = X^2$, $d_+(A) \geq d_-(A)$.*

Proof.   Consider a term $\prod a_{i\pi(i)} = \prod(\sum x_{ik} x_{k\pi(i)})$ in $d_-(A)$. A term in this equals $\prod x_{if(i)} x_{f(i)\pi(i)}$ where $f(i)$ is the k selected in the ith factor.

Case 1. Suppose f is not a permutation. Let $f(u) = f(v)$. Here π is odd. Then $\prod x_{if(i)} x_{f(i)\pi(i)} = \prod x_{if(i)} x_{f(i)\phi(i)}$ where $\phi(u) = \pi(v)$, $\phi(v) = \pi(u)$ and $\pi = \phi$ otherwise, since two terms in the product are interchanged. But ϕ differs from π by a transposition and is therefore even. So the latter term lies in $d_+(A)$.

Case 2. Consider all terms where f is a permutation. These are $\prod a_{if(i)} a_{f(i)\pi(i)}$ where π is odd for terms in $d_-(A)$ and even in $d_+(A)$. This can be written $\prod a_{if(i)} \prod a_{ig(i)}$ where f, g range over all pairs of permutations such that in $d_+(A)$, f, g have the same si and else the have opposite signs. The inequality follows from

$$(\sum_{f \text{ odd}} \Pi \ a_{if(i)})^2 + (\sum_{g \text{ even}} \Pi \ a_{ig(i)})^2$$

$$\geq 2 (\sum \Pi \ a_{if(i)})(\sum \Pi \ a_{ig(i)})$$

$$= \sum_{fg \text{ odd}} \Pi \ a_{if(i)} \ \Pi \ a_{ig(i)} \qquad\qquad \Box$$

EXAMPLE 4.4.2. This matrix has no square root over any incline since $d_+(A) = 0$, $d_-(A) = 1$.

$$\begin{bmatrix} 0 & 0 & 1 \\ 0 & 1 & 0 \\ 1 & 0 & 0 \end{bmatrix}$$

PROPOSITION 4.4.2.   *For* $A \in M_2(7_3)$, $d_+(A) \geq d_-(A)$ *is also sufficient that* $X^2 = A$ *and in this case* A *lies in a 1-parameter semigroup. For X to lie in* 7_3 *the additional condition* $Tr \ (A) \geq a_{12}{}^2 + a_{21}{}^2$ *is necessary and sufficient. Here* $Tr \ (A)$ *denotes the trace of* A.

PROOF. Let

$$X = \begin{bmatrix} a & b \\ c & d \end{bmatrix}$$

By symmetry under transpose and conjugation we may assume $b \geq c$, $a \geq d$. Then

$$X^2 = \begin{bmatrix} a^2 + bc & ba \\ ca & d^2 + bc \end{bmatrix}$$

By the same symmetries assume $a_{11} \geq a_{22}$, $a_{12} \geq a_{21}$. By assumption $a_{11}a_{22} \geq a_{12}a_{21}$. Let $a = \sqrt{a_{11}}$, $b = a_{12}/a$, $c = a_{21}/a$, and $d = \sqrt{a_{22}}$. Then $bc = a_{12}a_{21}/a_{11} \leq a_{22}$. Therefore $X^2 = A$ holds.

However X will have all entries at most 1 if and only if $b \leq 1$ since $c \leq b$. This will make X belong to

$M_2(7_3)$. Conversely $a^2 + bc + d^2 + bc \geq b^2a^2 + b^2d^2$ since $a^2 + d^2 \geq b^2(a^2 + d^2)$. □

EXAMPLE 4.4.3. This matrix has a square root:

$$\begin{bmatrix} 0.6 & 0.1 \\ 0.8 & 0.6 \end{bmatrix}$$

THEOREM 4.4.3. *For any i, j, t and integer, if $7=7_3$*

$$a_{ij}(t) = \sum_k a_{ik}(0) a_{kk}(\tfrac{t}{g})^g a_{kj}(0)$$

Proof. From the fact that $a_{ij}(t) = \sum a_{ii_1}(t_1) \cdots a_{i_kj}(t_{k+1})$ whenever $\sum t_i = t$ (because $A(t) = A(t_1) \cdots A(t_{k+1})$), it follows that $a_{ij}(t)$ is greater than or equal to the right hand side.

Regard $A(t)$ as B^r where $B = A(\tfrac{t}{r})$ and for some w, $r = wn! + 3n + 2n^3$. Take a term in the expansion of $A = B^r$ as $b_{ii_1} \cdots b_{i_{r-1}j}$. This term consists of cycles together with at most n factors (delete cycles until no more cycles can be deleted if more than n factors exist some one of $i \ldots j$ will occur twice giving a cycle). Among the cycles occurring let $Z = z_{kk_1} \cdots z_{k_{s-1}k}$ be the one for which the quantity $(z_{kk_1} \cdots z_{k_{s-1}k})^{1/s}$ is a maximum. Then if we delete any combination of other cycles before or after Z of total length L and insert a number of copies of Z of total length L the term will not be reduced. We do this as far as possible. For any length d there will remain at most $s - 1$ cycles of length d else we could eliminate s of them and replace with d copies of Z. So the total length of cycles before or after Z other than Z is at most $\sum (s - 1)d < n^3$. This gives at most $2n^3$ such factors. Therefore at least $wn!/s$ copies of

Z give a product $a_{kj}(m)$ for some p, m where $p + m \leq t(3n + 2n^3)/r$. The product of $wn!/s$ copies of Z can be divided into g groups if $g|w$. The total length M is at least $(1 - \frac{3n + 2n^3}{r})t$. Their product is at most $a_{kk}(\frac{M}{g})^g$. Therefore $a_{ij}(t) \leq \sum_k a_{ik}(p) a_{kk}(\frac{M}{g})^g a_{kj}(m)$. But as $w \to \infty$, p, $m \to 0$ and $\frac{M}{g} \to \frac{t}{g}$. □

THEOREM 4.4.4. *Let* $u_k(t) = \lim a_{kk}(\frac{t}{n!})^{n!}$. *Then*

$$a_{ij}(t) = \sum_k a_{ik}(0) a_{kj}(0) u_k(1)^t.$$

Proof. Since $a_{kk}(\frac{t}{g})^g \leq a_{kk}(t)$ because $A(\frac{t}{g})^g = A(t)$, any sequence $a_{kk}(\frac{t}{n_i})^{n_i}$ is decreasing provided $n_1 | n_2$, $n_2 | n_3$, ... Therefore such sequences being nonnegative have limits. Consider two sequences n_i, m_i where $n_i | n_{i+1}$, $m_i | m_{i+1}$ and every integer divides n_s, m_s for sufficiently large s. Then each n_i divides some m_j so that $a_{kk}(\frac{t}{n_i})^{n_i} \geq a_{kk}(\frac{t}{m_j})^{m_j}$. So the limit for n_i is greater than or equal to the limit for m_j. By symmetry the two are equal, and all equal the limit for $n!$.

Since $\lim a_{kk}(\frac{t}{n!})^{n!} = \lim a_{kk}(\frac{pt}{mn!})^{mn!/p} =$

$$= \left(\lim a_{kk}((\frac{pt}{m})(\frac{1}{n!}))^{n!}\right)^{m/p}$$

we have $u_k(t) = u_k(\frac{pt}{m})^{m/p}$ for integer p, m. So for t rational $u_k(t) = u_k(1)^t$. The equation $a_{ij}(t) = \sum_k a_{ik}(0) a_{kj}(0) u_k(t)$ holds simply by taking limits for $g = n!$ as $n \to \infty$, in the formula of the last theorem. Therefore for rational t the formula of this theorem holds. So it holds for all t by continuity. □

Idempotents such as $A(0)$ have a special form over

7_3.

THEOREM 4.4.5. *Any idempotent over 7_3 has the block form*

$$P \begin{bmatrix} A & AC \\ BA & BAC \end{bmatrix} P^T$$

where all main diagonal entries in A are 1.

Proof. We first show any row A_{k*} without a main diagonal 1 entry is dependent on rows having main diagonal ones $A_{k*} = \Sigma \, a_{kj} A_{j*}$. By repeatedly substituting we obtain $A_{k*} = \underset{S}{\Sigma} \, a_{ki_1} a_{i_1 i_2} \cdots a_{i_m j} A_{j*} + \underset{S^c}{\Sigma} \, a_{kj} A_{j*}$ where S is the set of rows with no main diagonal ones. It will be enough to show that the first set of terms converge to zero. We may assume none of A_{k*}, $A_{i_1 *}$, $A_{i_2 *}$, $\cdots$ has a main diagonal one or else we absorb these terms in the latter ones. Therefore every cycle involving k, i_1, $\cdots$, i_m has product less than or one and so all products of cycles converge to zero with m. So the entire term $a_{ki_1} a_{i_1 i_2} \cdots a_{i_m j}$ which consists of at most $n - 1$ terms after cycles are deleted, converges to zero uniformly in m. This proves the claim. Now choose p so that the first r rows are those with main diagonal ones. (The same holds for columns.) Then the last $n - r$ rows are dependent by a matrix B. The last $n - r$ columns are dependent by a matrix C. So we have the form

$$\begin{bmatrix} A & AC \\ BA & BAC \end{bmatrix} \qquad \square$$

EXAMPLE 4.4.4.

$$\begin{bmatrix} 1 & 0.2 & 0.3 & 0.5 \\ 0.3 & 1 & 0.09 & 0.5 \\ 0.5 & 0.1 & 0.15 & 0.25 \\ 0.3 & 0.2 & 0.09 & 0.15 \end{bmatrix}$$

THEOREM 4.4.6.   *There exists $\varepsilon > 0$ such that if $t < \varepsilon$ then $a_{ij}(t) = a_{ij}(0)v_{ij}^{t}$ for some v_{ij}.*

   Proof.   We have $a_{ij}(t) = \Sigma \, a_{ik}(0)a_{kj}(0)u_{k}(1)^{t}$. Delete terms with $u_{k}(1) = 0$. Then there exists ε such that for $t < \varepsilon$, any $u_{k}(1)^{t}$ is closer to 1 then any $(a_{ik}(0)a_{kj}(0))/a_{ij}(0)$ which is less than 1. It follows that we can ignore terms where $a_{ik}(0)a_{kj}(0) < a_{ij}(0)$. Let U be the set of remaining terms. Then $a_{ij}(t) = a_{ij}(0) \, \sup_{U} \, \{u_{k}(1)^{t}\}$. But if we let $v_{ij} = \sup_{U} u_{k}(1)$ the theorem follows.   □

EXAMPLE 4.4.5.

$$\begin{bmatrix} 0.6 & 0.8 \\ 0.2 & 0.6 \end{bmatrix}^2 = \begin{bmatrix} 0.36 & 0.48 \\ 0.12 & 0.36 \end{bmatrix}$$

So the latter matrix has a square root but does not lie in a 1-parameter semigroup over 7_3.

THEOREM 4.4.7.   *In order that for small t, $a_{ij}v_{ij}^{t}$ give a 1-parameter semigroup it is necessary and sufficient that if $a_{ij} > 0$ then $v_{ij} \geq \sup_{T} \{v_{ik}, \, v_{kj}\}$ and for some $k \in T$, $v_{ij} = v_{ik} = v_{kj} = v_{kk}$ and $a_{kk} = 1$ where $T = \{k: a_{ij} = a_{ik}a_{kj}\}$ and that (a_{ij}) be idempotent. Here we assume that if $a_{ij} = 0$, then $v_{ij} = 1$. Moreover, if $a_{ij} = a_{ii_1} \cdots a_{i_kj} > 0$ then $v_{ik} \geq v_{i_r i_{r+1}}$ for all r.*

   Proof.   Since $A(0)^2 = A(0 + 0) = A(0)$, A must be idempotent.

   In order to obtain $a_{kk} = 1$, we consider an n-fold product $A(\Sigma \, t_i) = \Pi \, A(t_i)$, $a_{ij}v_{ij}^{\Sigma \, t_i} = \Sigma \, a_{ii_1} \cdots a_{i_{n-1}j}v_{ii_1}^{t_1} \cdots v_{i_{n-1}j}^{t_n}$. This implies that if $a_{ij} = a_{ii_1} \cdots a_{i_{n-1}j}$ that $v_{ij} \geq v_{i_r i_{r+1}}$. Moreover, equality

must hold in some case. Since there at least n subscripts
i, i_1, i_2, ... , j there must be a cycle, and we can
write $a_{ij} = a_{ir} a_{rr} a_{rj}$ where r is any repeated subscript.
Moreover, we have $v_{ir} \geq v_{ii_1} = v_{ij}$ and $v_{ij} \geq v_{ir}$ and
$v_{rr} = v_{ij}$ so $v_{ir} = v_{ij}$. And $v_{rj} = v_{ij}$. So we can take
$a_{rr} = 1$ since $a_{ij} = a_{ir} a_{rr} a_{rj} \geq a_{ir} a_{rj}$.

Sufficiency can readily be checked from $a_{ij} v_{ij}^{s+t} =$
$\sum a_{ik} v_{ik}^s a_{kj} v_{kj}^t$. □

COROLLARY 4.4.8.   *For $a_{ij} > 0$, $v_{ij} = \sup\limits_{k \in D} \{v_{kk}\}$
where $D = \{k: a_{ik} a_{kj} = a_{ij}$ and $a_{kk} = 1\}$. Conversely,
this condition is sufficient.*

Proof.   Suppose $a_{ij} = a_{ik} a_{kk} a_{kj} > 0$ and $a_{kk} = 1$.
Then for a_{ik}, a_{kj}, $k \in T$ so $v_{ik} \geq v_{kk}$ and $v_{kj} \geq v_{kk}$.
So $v_{ij} \geq v_{kk}$. By the above theorem equality must hold
in some case, $v_{kk} = v_{ik} = v_{kj} = v_{ij}$ where $k \in D$.

Suppose this condition holds. Then $v_{ij} = v_{kk}$ and
$v_{ik} \geq v_{kk}$ and $v_{kj} \geq v_{kk}$ for some $k \in D$. It will suffice
to show that $v_{ij} \geq v_{ik}$ and $v_{ij} \geq v_{kj}$ for $k \in T$. Let
$v_{ik} = v_{bb}$ where $b \in D$ so $a_{ib} a_{bk} = a_{ik}$, $a_{bb} = 1$. Then
$a_{ib} a_{bj} \geq a_{ib} a_{bk} a_{kj} \geq a_{ik} a_{kj} = a_{ij}$ so $a_{ib} a_{bj} = a_{ij}$.
Therefore $v_{ij} \geq v_{bb}$. So $v_{ij} \geq v_{ik}$. By symmetry $v_{ij} \geq v_{kj}$. □

4.5 INCLINE STRUCTURES ON AN N-DIMENSIONAL CUBE

Here we will obtain a couple of results generalizing
those of Section 4.2 to n-dimensions. There exist
many distinct incline structures on $[0, 1]^n$ which are
subinclines of the product structure. We do not know
whether every incline structure with 0, 1 is such a
subincline. The fact that R^n has a unique commutative
topological group structure supports this view.

PROPOSITION 4.5.1.   *For any continuous additive operation on $[0, 1]^n$ which forms an incline with a multiplication of the form $(f_1^{-1}(f_1(x_1)f_1(y_1)), \ldots, f_n^{-1}(f_n(x_n)f_n(y_n)))$ where f_n is a homeomorphism of $[0, 1]$ to itself, we have the relations $(x + y)^n = x^n + y^n$ and more generally $\wedge M_i \leq \vee N_i$ where N_i, M_i are monomials and N_i is obtained from M_i by replacing one or more copies of a variable y_i by y_{i+1} (or y_n to y_1).*

Proof.   It suffices to show that the partial order must refine the standard partial order on $[0, 1]^n$ since these relations hold coordinatewise in it, since in each coordinate we have a linearly ordered incline.

But this is true since if $(u_1, \ldots, u_n) \leq (v_1, \ldots, v_n)$ then $(u_1, \ldots, u_n) = (v_1, \ldots, v_n)(w_1, \ldots, w_n)$ for some $(w_1, \ldots, w_n) \in 7$. $\square$

PROPOSITION 4.5.2.   *Let 7 be an incline structure on $[0, 1]^n$ having the usual addition and having multiplication a function of the $f(f^{-1}(a)f^{-1}(b))$ where f is a regular diffeomorphism from $[0, 1] \times [0, 1]$ to itself with f, f^{-1} preserving order. Then 7 is equivalent to the product structure on $[0, 1]^n$.*

Proof.   For f, f^{-1} to preserve order, they must be separately monotone in each variable. Thus their Jacobian matrices are nonnegative. But a nonnegative matrix having a nonnegative inverse has nonzero entries located only in entries $(i, p(i))$ for a permutation p by Theorem 2.2.3. Therefore after coordinates have been rearranged f has the form of Proposition 4.5.1. $\square$

4.6 ANALYSIS OVER AN INCLINE

In this section we consider derivative, measure, power series, quadratic form, and integral equations over 7.

For the incline R^+, inf $\{x, y\}$, $x + y$ every convex function can be considered rational over the incline. First $x + \ldots + x = nx$. And $nx = my$ yields $\frac{n}{m}x$. This gives $ax + b$.

THEOREM 4.6.1.   *Any convex downwards increasing function f from R^+ to R^+ is representable as an infinit series $\wedge$ $(a(r) + b(r)x)$ where $a(r) + b(r)x = y$ are all the lines bounding its graph.*

Proof.   A computation.   □

DEFINITION 4.6.1.   The *derivative* of a sum $\sum\limits_{i=0}^{\infty} a_i x^{b_i}$ is $\sum\limits_{i=0}^{\infty} a_i x^{b_i - 1}$ for $b_i > 0$.

PROPOSITION 4.6.2.   *The derivative is the unique operator on series $\sum a_i x^{b_i}$ satisfying $D1 = 0$, $Dx = 1$, $D \sum f_n = \sum D f_n$, $Dfg = fDg + gDf$ where Df denotes the derivative of f.*

There exists a Taylor series expansion for every power series $f(x) = \sum a_n x^n$. In fact $f(a + h) = \sum h^i \cdot f^{(i)}(a)$ as follows by expanding the series termwise. All infinite sums and products over an incline on $[0, 1]$ converge by monotonicity.

DEFINITION 4.6.2.   A *measure into a compact continuous incline 7* on $[0, 1]$ is a function from a set F of subsets of a set S into **7** such that (i) $m(\emptyset) = 0$, (ii) if $A_i \in F$ so does $\cup A_i$ and $m(\cup A_i) = \sum m(A_i)$.

The property (ii) is implied by the similar propert for disjoint unions. Note that this is not the same as our earlier definition of an invariant measure on **7**

EXAMPLE 4.6.1.   For a function $f(x)$, let $m(A) =$ $\sup\limits_{x \varepsilon A} \{f(x)\}$.

PROPOSITION 4.6.3.   *A measure with values in $\{0, 1\}$ is equivalent to a family F of subsets (for the sets of measure 0) such that (i) if $A \varepsilon F$ and $B \subset A$ then $B \varepsilon F$, (ii) if $A_i \varepsilon F$ so does $\bigcup\limits_{i=1}^{\infty} A_i$, and (iii) $\emptyset \varepsilon F$.*

EXAMPLE 4.6.2.   We can take F to be the family of nonempty subsets, countable sets, or sets which are countable unions of nowhere dense subsets. The latter gives Baire category. We can also take sets of Lebesgue measure 0.

EXAMPLE 4.6.3.   Consider coverings of A by n intervals of minimum length L_n. We can define F by $L_n = 0(1/n^2)$.

DEFINITION 4.6.3.   A function f is *measurable* if and only if $f^{-1}(B)$ is a measurable set for all intervals $B = [a, b]$.

EXAMPLE 4.6.4.   A constant function is measurable.

DEFINITION 4.6.4. $\int f(x)\ dm = \sup \Sigma\ a_i\ m(S_i)$ over all simple functions $\Sigma\ a_i \chi_{G_i}$ less than or equal to $f(x)$. Here $\chi_G(x) = 1$ if $x \varepsilon G$, 0 otherwise.

PROPOSITION 4.6.4.   *If $m(S) = 1$ for all nonempty S, then $\int f(x)\ dm = \sup\limits_{x} f(x)$, the maximum of $f(x)$ if it has one.*

As with Lebesgue integrals this quantity $\Sigma\ a_i m(S)$ is independent of the particular expression for a simple function.

THEOREM 4.6.5. $\int c\, f(x)\ dm = c\int f(x)\ dm$

$\int (f(x) + g(x))\ dm = \int f(x)\ dm + \int g(x)\ dm$

DEFINITION 4.6.5.   A *quadratic form* over 7^n is a function $F\colon 7^n \to 7$ such that $F((x_1,\ \ldots\ ,\ x_n)) = \Sigma\, a_{ij}x_i x_j$ for some a_{ij}.

EXAMPLE 4.6.5. $x_2 x_3 + 0.2 x_1 x_3$.

We have in matrix form $F(x) = xAx^T$. Since $xAx^T = xA^T x^T = x(A + A^T)A^T$, we may assume A is symmetric.

THEOREM 4.6.6.   *A function F satisfies $F(ax) = a^2 F(x)$ and $F(x + y) = F(x) + F(y) + b(x, y)$ for a bilinear function b if and only if it is a quadratic form.*

THEOREM 4.6.7.   *Every quadratic form over 7_3 can be replaced by one in which $a_{ij} = 0$ if $a_{ij} \le \sqrt{a_{ii}}\sqrt{a_{jj}}$, $i \ne j$. Otherwise coefficients are unique.*

An integral equation over an incline on $[0, 1]$ (or over any semiring structure on R with additive operation sup) takes this form: $\sup_x K(x, y)f(x) = g(y)$.

PROPOSITION 4.6.8.   *If an integral equation has a solution, its unique maximal solution is $f(x) = \inf_y \{g(y)/K(x, y)\}$.*

4.7 HILBERT SPACES OVER AN INCLINE

In this section we consider vectors of infinite

dimension over an incline 7. Unless otherwise specified
we assume 7 is isomorphic to 7_3. We characterize the
weak topology in which linear functionals are continuous, and characterize the following classes of linear
functionals (i) finitely additive, (ii) countably
additive, (iii) continuous in the product topology,
and (iv) continuous in the weak topology. We prove a
theorem on extension of a linear functional defined on
a subspace which applies also in the finite case. This
is a little like the Hahn-Banach theorem. We show that
a 1-1 onto mapping of compact metric topologies on an
incline Hilbert space must be continuous.

Then we turn to eigenvectors and eigenvalues. Invariant subspaces always exist but in the infinite
dimensional case, eigenvalues may not. For indecomposable matrices the only possible eigenvalue can be
calculated.

Let H denote the countably infinite product of 7
with itself, consisting of all sequences a_1, a_2, $\ldots$
from 7. A subspace of H is a subset closed under sums
and products from 7.

DEFINITION 4.7.1.   The *weak topology* on H is the
topology in which $f_n \to g$ if and only if $a \cdot f_n \to a \cdot g$
for all vectors a.

For brevity, let e_i denote the vector with a 1 in
the ith component and zero elsewhere.

EXAMPLE 4.7.1.   In the product topology $e_i \to 0$ but
in the weak topology $e_i \not\to 0$ since $e_i \cdot (1, 1, \ldots, 1)$
$= 1$.

The weak topology is actually stronger than the product topology since $e_i \cdot f_n \to e_i \cdot g$ if and only if f_n converges to g coordinatewise, that is in the product topology.

There are several topologies on $H = \overset{\infty}{\underset{1}{\times}} 7_3$. The product topology is compact and metric with metric $\Sigma (x_n - y_n)^2/n^2$ in which the operations are over the real numbers.

DEFINITION 4.7.2.   The *product topology* is the smallest topology in which each coordinate is continuous. This means that $f_n \to f$ if and only if $(f_n)_i \to f_i$ for integer i.

EXAMPLE 4.7.2.   The sequence e_i converges to 0 in the product topology but is nonconvergent in the weak topology since $(\overset{\infty}{\underset{1}{\Sigma}} e_i)e_i = 1$, yet $e_i e_j = 0$ for $i < j$.

DEFINITION 4.7.3.   A sequence of vectors f<n> *converges semiuniformly* to f if (i) f<n> $\to$ f in each coordinate and (ii) for every $\varepsilon > 0$ there exists N such that for $n < N$, $f<n>(1 - \varepsilon) \le f + \varepsilon(1, 1, \ldots , 1)$.

THEOREM 4.7.1.   *Convergence in the weak topology is equivalent to semiuniform convergence. If f<n> $\to$ f pointwise then for any a, lim a $\cdot$ f<n> $\ge$ a $\cdot$ f.*

Proof.   Since lim a $\cdot$ f<n> $\ge a_n f_n$ the last statement holds. If f<n> $\to$ f in the weak topology; (i) holds (let a have only one 1). If (ii) fails let $f<n>_m (1 - \varepsilon) > f_m + \varepsilon(1, 1, \ldots , 1)$ for some fixed ε for arbitrarily large n. Take a subsequence with m increasing. For a vector with $c_m = \varepsilon/(1 - \varepsilon)f<n>_m$ and $c_i = 0$ for other i we have a contradiction.   □

By a *linear functional* on a subspace V of H we mean
a homomorphism $f: V \to 7$, i.e., $f(ax) = af(x)$, $f(x + y)$
$= f(x) + f(y)$. If $f\left(\sum_{n=1}^{\infty} x_n\right) = \sum_{n=1}^{\infty} f(x_n)$, we call f
countably additive.

THEOREM 4.7.2.   *A linear functional on H is continuous in the weak topology if and only if it is countably additive. Both are equivalent to representability
as $\sum_i a_i x_i$.*

THEOREM 4.7.3.   *A linear functional on H is continuous in the product topology if and only if it can be
represented as $a \cdot x$ where $a_i \to 0$.*

   Proof.   Let $f = a \cdot x$, $a_i \to 0$ and $x{<}n{>} \to x$ in the
product topology.   By the above theorem, $\lim a \cdot x{<}n{>}$
$\geq a \cdot x$.   Suppose $\lim a \cdot x{<}n{>} = b > a \cdot x$.   Then for
all N, $\varepsilon > 0$ there exist $n > N$, m with $a_m x{<}n{>}_m > b - \varepsilon$.
But m is bounded since $b > 0$ and $a_m \to 0$.   So we may
assume it is constant by taking a subsequence.   Thus
$a \cdot x \geq \lim a_m x{<}n{>}_m \geq b - \varepsilon$.   This is a contradiction.
   Conversely, the product topology is weaker than the
weak topology, so by the last theorem a functional
continuous in either topology has the form $a \cdot x$.   If
a_i did not converge to zero, then $a \cdot e_i$ would not
converge to zero.   ☐

   We next consider the problem of extending a linear
functional defined on part of a subspace.   Firstly the
subspace should contain the vector with all ones.

EXAMPLE 4.7.3.   The function taking $(0.4, 1)$ to 1
and $(0.2, 1)$ to 0.5 is linear on the subspace they
generate but does not extend to the two-dimensional
vector space.

EXAMPLE 4.7.4. For any vector a the function $\overline{\lim}_i a_i v_i$ is finitely additive but not countably additive.

Secondly, there is a problem in the infinite case. Let $v = (1, 1, \ldots)$, $w = (1, \frac{1}{2}, \frac{1}{3}, \ldots)$, $f(v) = 1$, $f(w) = 0$. Then there is no vector a such that $ax = f(x)$ since $aw = 0$ implies $a = 0$.

THEOREM 4.7.4. *Let f be a linear functional on $V \subset H$ for a finitely generated subspace V containing $(1, 1, \ldots)$. Then there exists $a \varepsilon H$ such that $f(v) = av$ for all $v \varepsilon V$ if f is positive on all generators.*

Proof. We will take a as large as possible such that $a \cdot v \leq f(v)$ on a generating set. Let $v\langle 1 \rangle = (1, 1, \ldots)$, $v\langle 2 \rangle, \ldots, v\langle n \rangle$ span V. Let $a_i = \inf_j \{f_j v_{ji}^{-1}\} \leq 1$ where $v_{ji} = v\langle j \rangle_i$, $f_j = f(v\langle j \rangle)$, over j such that $v_{ji} \neq 0$. This guarantees $a \cdot v\langle j \rangle \leq f(v\langle j \rangle)$ since $a_i v_{ji} \leq f_j$. Therefore $a \cdot v \leq f(v)$ for $v \varepsilon V$.

Suppose equality fails. Then it fails for some $v\langle k \rangle$, $\varepsilon > 0$ and $a \cdot v\langle k \rangle + \varepsilon < f(v\langle k \rangle)$. We have $a_i = f_{j(i)} v_{j(i)i}^{-1}$ for some $j(i)$ (and $v_{j(i)i} \neq 0$).

The numbers $v_{j(i)i}^{-1}$ are bounded by $f_{j(i)}^{-1}$. Multiply by $\alpha = \inf_j \{f_j\}$. Therefore, we get

$$f(v\langle k \rangle \alpha) > \sum_{i=1}^{\infty} f_{j(i)} v_{j(i)i}^{-1} v_{ki} \alpha + \varepsilon \alpha$$

The right hand side is

$$\sum_{i=1}^{\infty} f(v\langle j(i) \rangle) v_{j(i)i}^{-1} \alpha v_{ki} + \varepsilon \alpha$$

Let $c_j = \sum v_{ki} \alpha v_{j(i)i}^{-1} \leq 1$ for $\{i: v\langle j(i) \rangle = v\langle j \rangle\}$. Then the right hand side is $\varepsilon \alpha + \sum_j c_j f(v\langle j \rangle) = \varepsilon \alpha + f(\sum_j c_j v\langle j \rangle)$. But $\sum_j c_j v\langle j \rangle \geq v\langle k \rangle \alpha$ since its mth coordinate is $\sum_i v_{j(i)m} v_{ki} \alpha v_{j(i)i}^{-1} \geq v_{km} \alpha$. This is a contradiction. $\square$

For a subspace of a finite dimensional subspace containing (1, 1, ... , 1) extension always can be done since $f = \lim_{\varepsilon \to 0} (f(x) + (\varepsilon, \varepsilon, \ldots , \varepsilon) \cdot x)$. These functions are positive, therefore, representable, and a subsequence of vectors a representing them will converge by compactness. Over 7_1 this extension theorem is false: let $f(1, 1) = 1$, $f(1, 0.5) = 1$, $f(0, 0.5) = 0.5$.

A *directed set* is a poset in which for all x, y there exists z > x, y. A *net* is a function from a directed set to another set. The *limit of a net* x_α is z means that for any open set U containing v there exists β such that if $\alpha > \beta$, $x_\alpha \in U$.

THEOREM 4.7.5. *A function from H to 7 is linear if and only if it is a pointwise limit of a net of functions of the form* $a_\alpha \cdot x$.

Proof. Let f be a linear functional. Take as directed set the product of all finitely generated subspaces V containing (1, 1, ...) directed under inclusion, and (0, 1) directed by x < y.

For given V, ε, let f_{V_ε} be $f(x) + \varepsilon(1, 1, \ldots)x$ restricted to V. Then f_{V_ε} is representable for all V, ε by the last theorem. Their limit pointwise is f(x) for any x: choose V containing x then we have $f(x) + \varepsilon \sum x_i \to f(x)$.

The converse is also straightforward. □

The open mapping theorem of functional analysis states that *a 1-1 linear mapping from a complete vector space to another must be continuous if the image of the unit ball contains a neighborhood of the origin.*

Here the last condition is not significant, and it seems preferable to replace it by the stricter condition that the mapping be onto. Then it essentially says that a complete metric topology on a linear space is unique.

THEOREM 4.7.6.   *The product topology is the unique first countable compact Hausdorff topology on H under which the operations x + y and kx are continuous.*
   Proof.   This follows directly from Theorem 4.1.3. $\square$

THEOREM 4.7.7.   *A countably additive function M: H → H is 1-1 if and only if for all n, i there exists j such that* $\frac{1}{n} m_{ij} > \sup_{k} m_{kj}$. *Here* m_{ij} *is the j coordinat of* $e_i M$ *where* e_i *is a (0, 1)-vector whose unique 1 entr is in coordinate i.*
   Proof.   It is 1-1 if and only if whenever v > w, vM > wM, since if v ≠ w but vM = wM we have a contradiction with w, v + w. This holds if and only if vM > wM whenever v equals except in coordinate i and v_i > w_i, since we can find v > z > w where z has this form. This holds if and only if $ke_i M$ is never less than $\sum_{j \neq i} e_j M$ for all k ε **7**, k > 0. And that is equivalent to the statement of the theorem since $\frac{1}{n} e_i M$ must excee $\sum_{j \neq i} e_j M$ is some coordinate. $\square$

EXAMPLE 4.7.5.   Form matrices $M<k> = I + \frac{1}{k}J$. Let M be obtained by writing the M<k> one after another: [M<1>M<2> ... M<k>]. Then M is 1-1.

DEFINITION 4.7.4.   If vA = kv (or Av = kv) and v ≠ 0, we call v an *eigenvector* of A and k an *eigenvalue* of A.

EXAMPLE 4.7.6.   The function sending $e_i \to e_{i+1}$ has no eigenvalues.

EXAMPLE 4.7.7.   The function sending $e_i \to \frac{1}{2}e_i + e_{i+1}$ has no eigenvalues.

DEFINITION 4.7.5.   A matrix A is *indecomposable* if and only if there does not exist a nonempty set S such that $a_{ij} = 0$ for $i \in S$, $j \notin S$.

EXAMPLE 4.7.8.

$$\begin{bmatrix} 0 & 1 & 0 \\ 0 & 0 & 1 \\ 1 & 0 & 0 \end{bmatrix}$$

PROPOSITION 4.7.8.   *Indecomposability is equivalent to* $\sum_{n=1}^{\infty} A^n$ *having all entries positive if 7 has the property that $ab > 0$ if $a > 0$ and $b > 0$.*

Proof.   If A is decomposable, so is $\sum_{n=1}^{\infty} A^n$. If $\sum_{n=1}^{\infty} A^n$ has its (i, j)-entry zero, let $S = \{v: a_{i\,i(1)} \cdot a_{i(1)i(2)} \cdots a_{i(k)v} > 0 \text{ for some } k\}$.   □

THEOREM 4.7.9.   *Let A be an indecomposable matrix and for any i let $c = \sup (a_{ii}^{(n)})^{1/n}$. Then c is independent of i and is the only possible eigenvalue of A.*

Proof.   Straightforward.

For example, if A has a cycle whose product is 1, then 1 is its only possible eigenvalue.

THEOREM 4.7.10.   *If an eigenvector exists for indecomposable A for eigenvalue c, it lies in the row*

space of the matrix $\inf_k \sum_{k=1}^{\infty} (c^{-1}A)^n$. And for all r, s,
$c^{-n}a_{r,s}^{(n)}$ must be bounded. Conversely, any vector in
this row space which is bounded and nonzero is an eige
vector assuming the sum is finite.

Proof. Straightforward.

EXAMPLE 4.7.9. Let $a_{ij} = \frac{1}{b}$ where $b = \sqrt{(i-j)^2 + 1}$
Then $a_{ii} = 1$ so $c = 1$. The first row of $A^4 = (1, 1/\sqrt{2}$,
$1/2, 1/2\sqrt{2}, 1/4, 1/\sqrt{26}, \ldots)$ is an eigenvector (it
equals the first row of all higher powers).

PROPOSITION 4.7.11. *Every matrix has a nontrivial
invariant subspace.*

Proof. Take the image space or if this is cH for
some c, then A has an eigenvector with eigenvalue c
since $c^{-1}A$ is an automorphism from H to H. □

THEOREM 4.7.12. *The indecomposable matrix A has a
row eigenvector if and only if each row of $(c^{-1}A)^n$ is
bounded independently of n, and not all entries conver
to zero.*

Proof. Straightforward.

EXAMPLE 4.7.10. This matrix has an eigenvector wi
eigenvalue 1 but not all entries of powers are bounded
(1) $a_{ij} = \frac{1}{2}$, $j = i + 1$; (2) $a_{ij} = 1$, $j = i$; (3) $a_{ij} =$
$j + 1 = i$; and (4) $a_{ij} = 0$, otherwise.

THEOREM 4.7.13. *The indecomposable matrix A has a
nonzero eigenvector provided that one of the following
hold: (i) $a_{ij} \to 0$ as $i + j \to \infty$ in any way; (ii) A is
symmetric and has a largest entry.*

Proof. Straightforward.

EXAMPLE 4.7.11.   Let A be a symmetric matrix with
$a_{nn} = 1 - 1/\sqrt{n}$, $a_{n+1,n} = 0.1$, $a_{n,n+1} = 0.1$, $a_{ij} = 0$
for other i, j. Then c = 1. But each entry of A^n
converges to zero. So A has no eigenvectors.

4.8 OPEN PROBLEMS

1. Is every incline structure on $[0, 1] \times [0, 1]$
    linearly representable ? does it satisfy $xy \leq x^2 + y^2$?

2. If $x_n \to y$ and $z_n = \sup_{k>n-1} \{x_k\}$ must $y = \inf \{z_n\}$ for
    compact 7 ?

3. If $x_n \to y$ and $w_n = \inf_{k>n-1} \{x_k\}$ must $y = \sup \{z_n\}$ for
    compact 7 ?

4. Under what conditions is every finite dimensional
    compact path connected incline with {0, 1} homeo-
    morphic to $[0, 1]^n$ for some n ? Can such an
    incline be nonmodular as a lattice ?

5. Define a Lie algebra of a regular differentiable
    incline.

6. For a compact lattice is this true: for any open
    neighborhood of a given x there exists an open
    neighborhood U of x such that $\sup U \in V$?

Chapter 5

Asymptotic Forms

The *asymptotic form* refers to the sequence of powers of a matrix. If a semiring is finite or a union of finite semirings as the fuzzy algebra, then the sequence of powers becomes periodic from some point on. The least power which occurs infinitely often is called the *index* and its period of repetition, the *period*. For Boolean and fuzzy matrices an exact formula exists for the maximum index.

For general inclines subject to conditions of compactness and linear ordering groups of matrices can be reduced to essentially the fuzzy case. However, if powers are not periodic we must either replace equality by less than or equal to, or consider the limits of powers of a matrix. There exists a formula for the limit. In the final section we compute the group theoretical complexity of matrices over many inclines.

5.1 INDEX OF BOOLEAN AND FUZZY MATRICES

We begin with the following definition.

DEFINITION 5.1.1. The *index* of a semigroup element

a is the least power a^k such that $a^k = a^s$ for some s > k. The *period* is the least integer d such that $a^k = a^{k+d}$.

EXAMPLE 5.1.1.   The matrix below has index 3:

$$\begin{bmatrix} 0 & 0 & 0 \\ 1 & 0 & 0 \\ 0 & 1 & 0 \end{bmatrix}$$

EXAMPLE 5.1.2.   This matrix has period 2:

$$\begin{bmatrix} 0 & 1 & 1 \\ 1 & 0 & 0 \\ 0 & 0 & 0 \end{bmatrix}$$

The index and period exist only if the cyclic semigroup generated by a is finite. (This is always true for $M_n(\beta)$ and $M_n(7_1)$. However, related concepts are defined for arbitrary inclines.

DEFINITION 5.1.2.   The *order-index* of an element a in an ordered semigroup is the least power a^k such that $a^k \geq a^r$ for some r > k. The *order-period* is the least integer d such that $a^{u+d} \geq a^u$ for some u.

THEOREM 5.1.1.   *Any matrix over any incline 7 has finite order-index and order-period.*
    Proof.   This follows since in the sequence A, A^2, ... , A^r, ... of powers of A some element must be greater than a succeeding element since they lie in a finitely generated incline (all n^2-tuples generated by their entries).   □

EXAMPLE 5.1.3.   Over ([0, 1], sup {x, y}, xy) this matrix has order-index 2 but no index.

$$\begin{bmatrix} 0 & 0.5 \\ 0.5 & 0 \end{bmatrix}$$

PROPERTY 5.1.1.   *Every sequence a(i) of length k + d from $\underline{n}$ has a subsequence b(1), ... , b(2m) for some m such that Σ b(2i) - b(2i - 1) = d and b(2i) = b(2i - 1).*

THEOREM 5.1.2.   *If k, d have the above property then $A^{k+d} \leq A^k$ over any incline (even noncommutative) for A ε $M_n(7)$.*

Proof.   Take any term $a_{ii_1} \ldots a_{i_s j}$ in $a_{ij}^{(k+d)}$. Delete the cycles from $a_{i_{2k-1}}$ to $a_{i_{2k}}$. Then we have a subsequence of length k which is a term in $a_{ij}^{(k)}$. ☐

THEOREM 5.1.3.   *Let D be any distributive lattice with basis. Then any matrix over D has index at most $(n - 1)^2 + 1$.*

Proof.   The lattice D is contained in a Boolean algebra β which is a direct product of copies of β indexed on the given basis. ☐

This theorem applies in particular to the fuzzy algebra.

5.2 GROUPS IN $M_n(1)$ AND CONVERGENCE

In this section we show that groups in $M_n(7)$ are similar to groups in $M_n(D)$ for Type I incline (see Definition 3.2.1). Groups in $M_n(D)$ can be related to symmetric groups. All inclines in this section will be Type I.

EXAMPLE 5.2.1.   For the multiplicative structure on [0, 1] this generates a group of order 3:

$$\begin{bmatrix} 0.2 & 1 & 0.4 \\ 0.4 & 0.2 & 1 \\ 1 & 0.4 & 0.2 \end{bmatrix}$$

Its image in $\mathbb{D}$ is

$$\begin{bmatrix} 0 & 1 & 0 \\ 0 & 0 & 1 \\ 1 & 0 & 0 \end{bmatrix}$$

Recall that we earlier proved that in a group in $M_n(7)$ if $A \geq B$ then $A = B$.

THEOREM 5.2.1.   *In an incline of Type I let h be the endomorphism $7 \to \mathbb{D}$ sending x to $\lim x^t$, and $\mathbb{D}$ be the distributive lattice of idempotents.   The mapping $h: M_n(7) \to M_n(\mathbb{D})$ is 1-1 on each $\mathcal{H}$-class which is a group.*

Proof.   Let A belong in a group with idempotent E. Let A belong to the kernel, so that $h(A) = h(E)$.   Let $F = h(E)$, $A^t = E$.   Then $e_{ij} = \Sigma \ e_{ik}f_{kk}e_{kj}$ by a previou[s] result.   We have $\lim a_{kk}{}^n = f_{kk}$ since $h(A) = F$. Then $a_{ij}{}^{(s)} \geq \Sigma \ a_{ik}{}^{(t)}a_{kk}{}^{s-2t}a_{kj}{}^{(t)} \geq \Sigma \ a_{ik}{}^{(t)}e_{kk}$ $\cdot a_{kj}{}^{(t)} = a_{ij}{}^{(t)}$ where the last equality holds since $a^{(t)}$ is idempotent.   So for $s > 2t$, $A^s \geq A^t$.   Thus $A^s = A^t$.   It follows that $A = A^t = E$.   So the kernel is E. $\square$

THEOREM 5.2.2.   *Suppose a distributive lattice $\mathbb{D}$ has basis the union of k chains C_k.   Then there exists a lattice monomorphism from $\mathcal{L}$ to $\times C_k^0$, where C_k^0 consists of zero together with C_k.*

Proof.   We define a mapping from L to C_k^0 as follows.
For $x \in L$, let $f(x)$ be the join of all elements of C_k
which are less than or equal to x. Suppose $f(x) = a$,
$f(y) = b$, $a \le b$. Then $x + y \ge b$ so $f(x + y) \ge b$. But
if $x + y \ge c$ since c is a basis element either $x \ge c$
or $y \ge c$ so $c \le b$. This proves $f(x + y) = f(x) + f(y)$.

We have $x \wedge y \ge a \wedge b = a$. So $f(x \wedge y) \ge a = f(x)$
$\wedge f(y)$. But suppose $x \wedge y \ge c \not\le a$. Then $x \ge c$ so
$f(x) \ge c$. This is false. The cases when $f(x)$ or $f(y)$
$= 0$ can be treated similarly. Therefore, we have a
lattice homomorphism.

It remains to show the combination of these mappings
is 1-1. Suppose $x \not\le y$. For some basis element b,
$b \le x$ and $b \not\le y$. And $b \in C_k$ for some k. Hence $f(x)$
$\ge b$ but $f(y) < b$.   □

THEOREM 5.2.3.   *If D is a linearly ordered distri-*
butive lattice then all groups in $M_n(D)$ are isomorphic
to subgroups of the symmetric group of degree n.

Proof.   Consider an H-class with idempotent E.   Let
B be in this H-class. All members of the H-class have
the same row space R(E). Multiplication $v \to vB$ is an
isomorphism from the row space to itself. Therefore,
it sends the unique standard basis to itself, so it
permutes the standard basis. This represents the H-
class by permutations. If the mappings $v \to vB$ and
$c \to vC$ are equal for all v then $B = EB = EC = C$.   So
the mapping is 1-1.   □

EXAMPLE 5.2.2.   This matrix has period 3 over the
fuzzy algebra:

$$\begin{bmatrix} 0.5 & 1 & 0.5 \\ 0.5 & 0.5 & 1 \\ 1 & 0.5 & 0.5 \end{bmatrix}$$

THEOREM 5.2.4.   *Suppose 7 is Type I and that $\mathcal{D}$ is finite, B is a basis for $\mathcal{D}$, and B is a union of k linearly ordered sets. Then $M_n(7)$ contains an isomorphic copy of SG_n^k where SG_n is the symmetric group on $\underline{n}$, and every group in $M_n(7)$ is isomorphic to a subsemigroup of SG_n^k.*

Proof.   By the previous result it suffices to prove this for $M_n(\mathcal{D})$.   Let $(a_1, \ldots, a_k)$ be an antichain of size k in B.   To every k-tuple of permutation matrices $P_1, \ldots, P_k$ we associate the element $\Sigma a_i P_i + \Sigma a_i a_j J$ This is a homomorphism since $a_i a_j P_i P_j < a_i a_j J$, so only products with the same a_i change.   The matrices P_i can be recovered by multiplication of the expression by a_i since a_i is a basis element.   So it is 1-1.

The converse follows from the preceding two theorems

$\square$

COROLLARY 5.2.5.   *Let c_n be the least common multip. of the elements of $\underline{n}$.   Then every element of a group in $M_n(7)$ has order dividing c_n.*

5.3 CONVERGENCE

We shall begin with the following theorem.

THEOREM 5.3.1.   *Let $h: 7 \to \mathcal{D}$ be the mapping which assigns to each $x \in 7$ the limit of x^n for a Type I incline 7 where $\mathcal{D}$ is the distributive lattice of idempotents.   For any matrix $A \in M_n(7)$, the matrix $h(A)$ has finite index and period.   If $h(A)$ has period 1 then A^m converges to the matrix $\sum_{v=1}^{n-1} A^u BA^v$ for any $u \geq n - 1$ where B is any idempotent power of $h(A)$.*

Proof.   We have $A \geq h(A)$ so for all sufficiently large s, $A^s \geq A^u h(A)^{s-u-v} A^v = A^u BA^v$ since B is idempotent.   It suffices to prove the reverse.   Consider any term T in the expansion of $a_{ij}^{(t)} = a_{ii_1} \ldots a_{i_r j}$.

We can express the sequence i, i_1, $\ldots$, i_r, j as a union of cycles and a path from i to j. Choose $t \geq n$ + $n^n nrc$. Then some cycle occurs at least to the rc power since there are at most n^n cycles having degree at most n, and the path has length less than n. Delete all factors except powers of this cycle Z and paths from i to k and k to j of length at most $n - 1$ where k occurs in Z. By rotating the cycle Z change the length of the first path to u. Delete cycles from the second path to shorten it. This gives a term $A^u h_1(A)^d A^v$ where $h_1(A) = (a_{ij})^r$ and $d \geq c - u - v$. Therefore $A^t \leq \Sigma A^u h_1(A)^d A^v$ and therefore $\overline{\lim} A^t \leq \Sigma A^u h(A)^d A^v$ where $d = c - u - v$. But for d at least equal to the index of $h(A)$ we have $h(A)^d = B$.   □

COROLLARY 5.3.2.   *Under the same hypotheses on* 7, *for $h(A)$ of period $p > 1$, for any r_0 the sequence of powers $A^{r_0 + tp}$ converges to $\Sigma A^j BA^k$ where $j = r_1 + up$ and $k = r_2 + vp$ whenever $r_1 + r_2 = r_0$ and B is an idempotent power of $h(A)$ and $u \geq n - 1$ and $v = 1, 2,$ $\ldots$, $n - 1$.*

Proof.   By the theorem applied to A^p we have A^{tp} converges to $A^{up} BA^{vp}$. Multiply on right and left by A^{r_1}, A^{r_2}.   □

EXAMPLE 5.3.1.   This matrix

$$\begin{bmatrix} 0 & 1 & 0 \\ 0 & 0 & 1 \\ 1 & \frac{1}{2} & 0 \end{bmatrix}$$

has B = I.   Therefore, it converges finitely.   Its
powers are:

$$A^2 = \begin{bmatrix} 0 & 0 & 1 \\ 1 & 0.5 & 0 \\ 0 & 1 & 0.5 \end{bmatrix}, A^4 = \begin{bmatrix} 0 & 1 & 0.5 \\ 0.5 & 0.25 & 1 \\ 1 & 0.5 & 0.25 \end{bmatrix}$$

$$A^8 = \begin{bmatrix} 0.5 & 0.25 & 1 \\ 1 & 0.5 & 0.25 \\ 0.25 & 1 & 0.5 \end{bmatrix}$$

The last equals any larger power congruent to it modu-
lo 3.

EXAMPLE 5.3.2.   Let a matrix A over [0, 1] have all
entries zero except in locations (j, j) and (i, i + 1).
Let the a_{jj}-entry be 1 and entries $a_{i,i+1}$ ε (0, 1).
Then b_{jj} = 1 and b_{rs} = 0 for (r, s) ≠ (j, j).   The
limit of the powers of A has (r, s)-entry $a_{r,r+1} a_{r+2,r+}$
$\cdots a_{j-1,j} a_{j,j+1} \cdots a_{s-1,s}$ for r ≤ j ≤ s and others of
its entries are zero.
     The matrix ABA has nonzero entries only in locations
(j, j), (j - 1, j), (j, j + 1), (j - 1, j + 1) as do
its powers.   They do not give the limit.

PROPOSITION 5.3.3.   *The order-period of a matrix M*
over a Type I incline is a multiple of period of h(M).
     Proof.   Let M, h(M) have periods d, d_1. If M^r ≥
M^{r+d} then $h(M)^r$ ≥ $h(M)^{r+d}$ and for r ≥ index of h(M),

equality must hold.   Thus d is a multiple of d_i.   □

EXAMPLE 5.3.3.   Equality may not hold:

$$\begin{bmatrix} 0 & 0.5 \\ 0.5 & 0 \end{bmatrix}$$

This has order-period 2, but its image has order-period 1.

THEOREM 5.3.4.   *A matrix A in $M_n(7)$ fits into exactly one of these classes (i) convergent to J, (ii) convergent to zero, (iii) convergent to an idempotent E, 0 < E < J, and (iv) powers A^{1+ud} converge to a matrix of period d > 1 for some case.   In every case A fits into the same category has h(A) does.*

Proof.   This result has already been proved except for the distinction between converging to J and converging to an idempotent E < J.   Since A > h(A), if h(A) converges to J then A must.   But if A converges to J then A^r must equal J for finite J since from some point on we have $A^r \geq A^{r+d}$.   And if A^r = J then $h(A)^r$ = h(J) = J.   □

EXAMPLE 5.3.4.   This matrix converges to J:

$$\begin{bmatrix} 0.1 & 1 & 0.5 \\ 0.2 & 0.3 & 1 \\ 1 & 1 & 0.4 \end{bmatrix}$$

This does not

$$\begin{bmatrix} 1 & 0.6 & 0.6 \\ 1 & 1 & 1 \\ 1 & 1 & 1 \end{bmatrix}$$

Shift equivalence (see Kim and Roush (1979), and Williams (1974)) is a relation generalizing conjugacy useful for matrices which are not over a field.

DEFINITION 5.3.1.   In any semigroup S, a ~ b if and only if there exist r, s ε S, and n $\geq$ 0 such that ra = br, as = sb, rs = b^n, sr = a^n.

The traces of all powers of a matrix are shift equivalence invariants over all commutative semirings.

PROPOSITION 5.3.5.   *Let a be an element of a semigroup S. If a^t is idempotent, then a ~ a^{t+1}.*
    Proof.   Let r = s = a^t.   □

THEOREM 5.3.6.   *Let a, b be the elements of a semigroup S. Let H_x denote {y: x H y}. Here H_x denotes Green's equivalence class. Let H_a, H_b be groups. Then a ~ b if and only if there exist elements x, x* such that x* is a Vagner inverse of x and a = xbx*, and b = x*ax.*
    Proof.   Let r = bx* = (x*ax)x* = x*xbx*xx* = x*xbx* = x*a.   Let s = xb = ax, n = 2.   □

These results settle shift equivalence over 7_1, 7_2. Over 7_3 we have a strong asymptotic form.

LEMMA 5.3.7.   *For given n there exists k such that for any path from i to j of length k + n! there exists a subpath of length k obtained by deleting cycles.*
    Proof.   Break a path of length k + n! into segments of length n + 1. Each segment has a cycle of some length 1, ... , n. Take nn! segments so that there are n! of some particular length.   □

THEOREM 5.3.8.   *For a matrix A over 7_3 for any sufficiently large k, $A^{k+n!} = C \odot A^k$ where C is the matrix whose (i, j)-entry is the $n!$ power of the value of the largest cycle occurring in a positive path from i to j. Here $\odot$ denotes an elementwise product.*

Proof.   By the above lemma for k sufficiently large $a_{ij}^{(k+n!)} \leq c_0 a_{ij}^{(k)}$ for some product c_0 of cycles occurring in a positive path from i to j. So $A^{k+n!} \leq C \odot A^k$.

For any length k path from i to j in which the maximum cycle occurs we can insert enough copies to multiply that by c_{ij} and have a path in $A^{k+n!}$. We will show that for k large such a path will have the largest value. This is true because the other paths except for n factors are made up of cycles of value at most the next lower value y and $n!$ so such a path has product at most y^{k-n}. And paths of the former type have values of the form $ac_{ij}^{(k-b)}$ for positive constants a, b. Eventually $ac_{ij}^{(k-b)} > y^{k-n}$. So $A^{k+n!} \geq C \odot A^k$.   □

COROLLARY 5.3.9.   *If $\sum\limits_{n=1}^{\infty} A^n$ has no zero entries there exists a constant $c > 0$ such that $A^{k+n!} = cA^k$ for all sufficiently large k.*

Then shift equivalence can be decided by shift equivalence of the matrices $c^{-k}A^k$ over the semiring $[0, \infty)$. These matrices have finite index and so we can apply Proposition 5.3.5 and Theorem 5.3.6.

5.4 SCHEIN RANK AND GROUP-THEORETIC COMPLEXITY

Group-theoretic complexity of finite semigroups is a concept introduced by John Rhodes and is important in both semigroup theory and automata theory.

DEFINITION 5.4.1.   Let S_1 and S_2 be semigroups and Y a homomorphism from S_1 into the endomorphism semigroup of S_2. Then the *semidirect product* $S_2 \underset{Y}{\times} S_1$ is the set $S_2 \times S_1$ with operation $(a, b)(c, d) = (aY(b)(bd)$. The notation $S_k \underset{Y_{k-1}}{\times} \cdots \times S_2 \underset{Y_2}{\times} S_1 \underset{Y_1}{}$ denotes

$$(\ldots((S_k \underset{Y_{k-1}}{\times} S_{k-1}) \underset{Y_{k-2}}{\times} S_{k-2}) \cdots \underset{Y_1}{\times} S_1).$$

EXAMPLE 5.4.1.   The semidirect product of Z_2 and Z where Z_2 acts on Z by multiplication by -1, consists of pairs $(\pm 1, n)$ where $n \in Z$ and products are $(a, b)($ $d) = (ac, bc + d)$. This is isomorphic to the group o linear transformations $ax + b$, $|a| = 1$, $b \in Z$.

EXAMPLE 5.4.2.   Let G be a group of permutations o $\underline{n}$. Let T be an arbitrary semigroup. Then G acts on the n-fold Cartesian product T^n by permuting the factors. The semidirect product in this case is called a *wreath product*. All semidirect products by a finit group are subsemigroups of wreath products.

DEFINITION 5.4.2.   The *group complexity* $\#_G(S)$ of a finite semigroup S is the least nonnegative integer k such that S is a homomorphic image of a subsemigroup of some semigroup $C_k \underset{Y_{k-1}}{\times} G_k \underset{Z_{k-1}}{\times} C_{k-1} \underset{Y_{k-2}}{\times} G_{k-1} \underset{Z_{k-2}}{\times}$ $\cdots C_1 \underset{Y_0}{\times} G_1 \underset{Z_0}{\times} C_0$ where the C_i are finite semigroup whose H-classes contain one element and the G_i are finite groups.

EXAMPLE 5.4.3.   A semigroup whose H-classes have only one element has complexity zero.

DEFINITION 5.4.3.   A *combinatorial semigroup* is one which contains no nontrivial finite groups.

EXAMPLE 5.4.4.   A finite incline 7 is combinatorial.

THEOREM 5.4.1.   *A finite semigroup is combinatorial if and only if its H-classes have at most one element if and only if its complexity is zero.*

Proof.   If the H-classes have at most one element, then there can be no nontrivial finite groups, since each lies in an H-class. Suppose H is an H-class having more than one element. Let $u, v \in H$. Let $xu = v$. Then by Green's Lemma multiplication by x sends H into H and is an isomorphism. Since the semigroup is finite some power x^e of x is idempotent. Then all powers x^m, $m \geq e$ form a group. But some power of x^e (which must equal x^e) is the identity on H. So $x^{e+1}u = v$. So $x^{e+1} \neq x^e$ and there is a nontrivial group.

A combinatorial semigroup S, by definition, has complexity 0 since we can take $G = S$. To prove the converse it suffices to note that any homomorphic image of a subsemigroup of a semigroup with no nontrivial groups has none. For subsemigroups this is immediate. Suppose S has no nontrivial groups and let $f: S \to T$ be an epimorphism. Suppose T has an H-class H_1 which is a nontrivial group. Let $f(x)$ be a nonidentity element of H_1. Let x^e be an idempotent power of x. Then powers x^i for $i \geq e$, form a group. And $f(x^{e+1}) = f(x)$ is nonidentity in H_1 so $x^{e+1} \neq x^e$. □

EXAMPLE 5.4.5.   Any nontrivial finite group has complexity 1.

John Rhodes has proved the following basic and important results on group-theoretic complexity.

AXIOM 5.4.1. *If S is a subdirect product of S_1, ..., S_k, then $\#_G(S) = \max\{\#_G(S_i)\}$.*

RESULT A. *Suppose S has a unique maximal J-class J which is regular. Let e be any idempotent of J. Then $\#_G(S) = \#_G(eSe)$. Here J denotes Green's equivalence class.*

RESULT B. *Let K be an ideal of S. Then $\#_G(S) \leq \#_G(S/K) + \#_G(K)$. Here "/" denotes the quotient semigroup.*

RESULT C. *If every R-class of S contains at most one idempotent, then $\#_G(S) \leq 1$. Here R denotes Green' equivalence class.*

RESULT D. *Let S be the semigroup of transformation on $\underline{n}$. Then $\#_G(S) = n - 1$.*

Details can be found in Allen (1971), Krohn, Rhodes and Tilson (1968), Krohn and Rhodes (1968), Rhodes (1966, 1967, 1968, 1971, 1974), Rhodes and Tilson (1971), Tilson (1971).

THEOREM 5.4.2. *A finite linearly ordered x-simple incline 7 with identity 1 consists of 1 together with elements y such that $ya < a$ for all $a > 0$.*

Proof. Let $a_0 = 1 > a_1 > a_2 > ... > a_n = 0$ be a finite linearly ordered x-simple incline. Then $a_{n-1} = x$ since x is a minimal nonzero element. For all $a > b$, there exists u such that $au = x$, $bu = 0$.

Let $b_n = 0$, and for $k > 0$, $b_k a_{k-1} = x$, $b_k a_k = 0$. Then $0 = b_0 < b_1 < b_2 < \ldots < b_n = 1$. Therefore $b_k = a_{n-k}$. Therefore $a_{k-1} a_{n-k} = x$, $a_k a_{n-k} = 0$. This property uniquely characterizes a_{n-k} since the b's are strictly linearly ordered. We have $a_1 a_i \leq a_{i+1}$ since $a_1 a_i a_{n-i-1} = a_1 x = 0$. But this implies the result since $ya \leq a_1 a$.

□

EXAMPLE 5.4.6.   The following is an x-simple linearly ordered incline which is not cyclic: $1 = a_0 > a_1 > a_2 > a_3 > a_4 = 0$ where $a_1 a_2 = a_3$, $a_1 a_3 = a_2 a_2 = 0$, $a_1^2 = a_3$.

THEOREM 5.4.3.   *Suppose a Type I incline 7 has no idempotents except 0, 1. Then the basis for any subspace over 7 is unique.*

Proof.   We first show all bases are standard. If $x = cx + \Sigma v_j$ then by substitution $x = c^2 x + \Sigma v_j = c^n x + \Sigma v_j$. So $x = \lim c^n x + \Sigma v_j$. If $c < 1$, $\lim c^n$ is an idempotent which must be zero.

Suppose $x = ax + bx$. Then $x = (a + b)^n x$ and if a, $b < 1$, $(a + b)^{2n} \leq a^n + b^n$ approaches zero. So a or b is 1. Now suppose we have two bases x_i, y_j where $x_i = \Sigma a_{ij} y_j$, $y_j = \Sigma b_{ji} x_i$. Then $x_i = \Sigma a_{ij} b_{jk} x_k$. And $x_i = \Sigma a_{ij} b_{ji} x_i$. And for some j, $a_{ij} b_{ji} = 1$. So $x_i = y_j$. Likewise every y_j equals some x_i. □

EXAMPLE 5.4.7.   This applies to 7_2, 7_3.

THEOREM 5.4.4.   *Suppose a matrix A over $M_n(7)$ has row rank n and there exists a unique basis for its row space. Then its Green's L-class consists of all matrices PA where P is a permutation matrix. If it is regular, its Green's R-class contains a unique idempotent.*

Proof.   Exercise.

EXAMPLE 5.4.8.   These two idempotents lie in the same R-class, having rank less than n:

$$\begin{bmatrix} 1 & 0 \\ 0 & 0 \end{bmatrix}, \qquad \begin{bmatrix} 1 & 1 \\ 0 & 0 \end{bmatrix}$$

DEFINITION 5.4.4.   The *Schein rank* of an n × n matrix over a semiring is the least k such that the matrix is the product of an n × k matrix and a k × n matrix.

EXAMPLE 5.4.9.   This matrix has Schein rank 3:

$$\begin{bmatrix} 1 & 1 & 0 & 0 \\ 1 & 1 & 1 & 0 \\ 0 & 1 & 1 & 1 \\ 0 & 0 & 1 & 1 \end{bmatrix}$$

LEMMA 5.4.5.   *If a matrix has Schein rank less than or equal to k, then it can be written as $AE_k B$ where E_k is the matrix whose (i, j)-entry is zero if $i \neq j$ or $i \leq k$ and is 1 otherwise.*

Proof.   Add zero columns and rows to the matrices of the definition.   □

EXAMPLE 5.4.10.   For the preceding example we have:

$$\begin{bmatrix} 1 & 0 & 0 \\ 1 & 1 & 0 \\ 0 & 1 & 1 \\ 0 & 0 & 1 \end{bmatrix} \begin{bmatrix} 1 & 0 & 0 \\ 0 & 1 & 0 \\ 0 & 0 & 1 \end{bmatrix} \begin{bmatrix} 1 & 1 & 0 & 0 \\ 0 & 1 & 1 & 0 \\ 0 & 0 & 1 & 1 \end{bmatrix}$$

PROPOSITION 5.4.6.   *Matrices of Schein rank at most k form an ideal. If a matrix has Schein rank k, its row and column rank are at least k.*

Proof.   Let A = BC. Then XAY = (XB)(CY) for B of size n × k, C of size k × n, X, Y, A of size n × n. And if A has row rank r let C be a matrix whose rows form a row basis for R(A). Then A = BC for some B. □

THEOREM 5.4.7.   *For any finite incline 7 which is a subdirect product of inclines whose only idempotents are 0, 1, the group-theoretic complexity of $M_n(7)$ is n - 1.*

Proof.   Exercise. Use all of Axiom 5.4.1, A, B, C, D. □

5.5 OPEN PROBLEMS

1.   Compute the maximum order-index of a member of $M_n(7)$.

2.   Find a way to determine the order-period.

3.   Find a criterion that the index of a matrix in $M_n(7)$ exist.

4.   For 3 × 3 matrices over any incline (even non-commutative) is $x^5 \geq x^{11}$? Then the order-index is at most 5.

5.   Is $\lim A^n = A^n B A^n$ where A converges ?

Chapter 6

Applications of Inclines

All applications of Boolean algebra and fuzzy algebra
are instances in which inclines can be applied. These
applications include automaton theory, design of swit-
ching circuits, logic of binary relations, medical
diagnosis, Markov chains, social choice, models of
organizations, information systems, and political systems, and clustering.

Here we outline additional applications involving
inclines other than Boolean algebras and fuzzy algebra.
We give models of political and social systems, and
the nervous system. We give applications to probable
reasoning, choice behavior, and finite state machines.
Inclines can be used to find least-cost paths in graphs
and for simple dynamic programming, as well as clustering. They have applications to matrix theory, in particular asymptotic properties of polynomial matrices.

6.1 LINEAR SYSTEMS

A number of types of systems in physical science can be approximated by x[i + 1] = x[i]A[i] + B[i]. In fact for any differential system x' = f(x, t) where x, f are vectors we have approximately x[i + h] = x[i] + f(x[0], i) + h(x[i] - x[0])J where J denotes here the Jacobian matrix. To be more specific such systems occur in the study of general damped vibrations in systems of electrical circuits involving resistance, inductance, and capacitance, and in the theory of molecular vibrations. More generally, almost all systems in classical mechanics can be expressed by linear or nonlinear differential equations. These can be approximated by the discrete systems mentioned.

The key distinction between systems involving an incline and the above-mentioned systems is that the additive operation (which is typically *sup* or *inf*) is idempotent. It is almost never the case for physical forces that we have quantities x, y to give an output sup {x, y}. A few exceptions occur in optics: for instance, the light obscured by A together with B equals the union of the light obscured by A and that by B. Reflections can also be seen as a phenomenon of this nature. For instance, the path of a reflected light ray is something like |x| = sup {x, -x}. As a third example, a change of phase, as from solid to liquid, may be explainable in this way: the resulting state is one or the other, according to, say, the inf of the potential energy of the two states.

In general systems involving an inf or sup are essentially cybernetic: some decision-making element chooses one or the other according to its design.

EXAMPLE 6.1.1.   We can consider the following con-
cerning the human nervous system. Each nerve cell i
has a state x_i in $[0, 1]$. Each cell exerts an influ-
ence $a_{ij} * x_j$ on x_i where a_{ij} is a constant measuring
the degree of affinity between the cells (or the memory
of past associations between them) and $x * y$ is some
incline operation such as xy. Then we may reasonably
assume the state of cell i is the supremum of the
influences which come to it, $x_i(t + 1) = \sup \{a_{ij} *$
$x_j(t)\}$. It may be useful also to consider dimensions
greater than 1 with the state being an element of
$[0, 1]^n$.

EXAMPLE 6.1.2.   We can consider the political system of a country i as influenced by the neighboring
countries, as an element x_i of $[0, 1]$. The state is
influenced by that of other countries x_j. The influ-
ence varies with the proximity of the other country
by a constant factor a_{ij}. Thus the state of country
i at time $t + 1$ might be $x_i(t + 1) = \sup \{a_{ij} x_i(t)\}$.

EXAMPLE 6.1.3.   In sociology a person's opinion or
adoption of a fad x_i may be measured on a $[0, 1]$ scale.
It may be influenced by those around him in such a way
that $x_i(t + 1) = \sup \{a_{ij} x_j(t)\}$ where a_{ij} represents
the degree to which person i is influenced by person j.

In 7_1 all elements are idempotent, in 7_2 all elements have finite index, and in 7_3 no elements are of
finite index. Also the operations can be compared:
$\inf \{x, y\} \geq xy \geq x + y - 1$. Unless otherwise stated
we will assume applications in this chapter are to 7_3.
However most of these can also be applied to 7_2.

EXAMPLE 6.1.4.   Consider four countries, (1) the
United States, (2) Japan, (3) Philippines, and (4) West
Germany.   Let human rights and freedom be measured on
a scale [0, 1] by 7_3.

	Present Degree		Influence			
			(1)	(2)	(3)	(4)
(1)	0.8	(1)	1	0.8	0.3	0.8
(2)	0.7	(2)	0.5	1	0.2	0.5
(3)	0.3	(3)	0.1	0.1	1	0.1
(4)	0.7	(4)	0.5	0.5	0.2	1

Then the matrix multiplied by the column vector equals
the column vector, so the system is stable. Moreover,
the matrix itself is idempotent.

In Cao (1982a, 1982b), systems over an incline are
applied to control theory of psychological phenomena.

6.2 APPLICATIONS IN PROBABLE REASONING, CHOICE, AND AUTOMATA

As a first method of using inclines in probable reasoning suppose each of a set of statements supports some
given conclusion C. If statement i supports C to a
degree a_i and one's confidence in statement i is p_i
then one's confidence in conclusion C can be taken as
being at least $\sup_i \{a_i p_i\}$. If in fact all data were
known to be independent a higher value could be obtained $1 - (1 - a_1 p_1)(1 - a_2 p_2) \cdots (1 - a_n p_n)$.
However for most values the lower value $\sup_i \{a_i p_i\}$
is a good approximation.

PROPOSITION 6.2.1.   *If x, y are independent and uniformly distributed in $[0, 1]$ the expected value of $1 - (1 - x)(1 - y) - \sup \{x, y\}$ is $1/12$.*

Proof.

$$2\int_o^1 \int_o^y ((1 - (1 - x)(1 - y)) - y) \, dx \, dy = 1/12.$$

$\square$

In general, $\sup \{\text{Prob } (p_i)\}$ can be used as an estimate of the probability of $p_1 \vee \ldots \vee p_n$. It is a lower bound and is exact if and only if some p_i implies all the rest.

Inclines can be used to represent binary relations in which there is a notion of degree as well as existence or nonexistence of the relationship, and in which the composition may have lesser degree than its factors.

Furthermore, inclines can be used to represent choice or decision behavior by a consumer or a mechanism governing some process. For example, a consumer is assumed to choose the bundle of goods giving him the highest utility among all those he can purchase. Therefore, if he can afford amount x_i of good i and the utility of this to him is a_i per unit, his utility is $\sup_i \{a_i x_i\}$.

If it is desired to have the quantity x_i as well as the utility enter the calculation, we may use a product structure. Then $\sup_i \{(a_i x_i, x_i)\}$ represents the utility and quantity chosen where we effectively order the pairs according to the first factor only. This can be achieved, for example by a lexicographic linear order or a partial order.

Inclines can be used to indicate quality as opposed to quantity, or order of magnitude, since sums preserve values. The sum of two items of quality x may be taken as quality x.

A cybernetic mechanism can govern a process by optimizing some quantity which is its goal. For instance, it can minimize the distance from an ideal state. If the quantity can be expressed as $a_i x_i$ for choice of i then again we have sup $\{a_i x_i\}$.

We can consider finite machines over an incline 7 in basically the same way as linear systems.

DEFINITION 6.2.1.   (Cao.) A *module* over 7 is a commutative semigroup M provided with a binary operation $7 \times M \to M$ satisfying (i) $a(x + y) = ax + ay$, (ii) $a(bx)$ = (ab)x, (iii) (a + b)x = ax + bx, (iv) 0 + x = x, and (v) $1x = x$.

EXAMPLE 6.2.1.   Let K be an ideal in 7. Then K is a module over 7.

DEFINITION 6.2.2.   A *free module* over 7 is a module isomorphic to a direct sum of copies of 7.

EXAMPLE 6.2.2.   The vector space V_n of all n-tuples of elements of 7 is a free module.

A free module is itself an incline.

PROPOSITION 6.2.2.   (Cao.) *A module over 7 is a semilattice. We have $ax + x = x$, $a0 = 0$, and if $x + y = 0$ then $x = y = 0$.*

     Proof. $x + x = (1 + 1)x = 1x = x$, $x + ax = (1 + a)x$ $= 1x = x$, $a0 = a0 + 0 = (a + 1)0 = (1)0 = 0$. The last result is true in any semilattice. □

THEOREM 6.2.3.   (Cao.) *A free module over integral 7 has a unique basis.*

Proof.   If there were two bases, the matrices A and B expressing each in terms of the other would satisfy AB = BA = I. By the previous theorem about invertible matrices over an integral incline A, B are permutation matrices.   □

The degree of a mapping from one free module to another is its algebraic degree in terms of coordinates: a linear map has degree 1, a bilinear or quadratic map has degree 2.

DEFINITION 6.2.3.   A *finite state machine* (S, X, ν, δ) over 7 consists of modules S, X, Z over 7 and mappings ν: S × X → S and δ: S × X → Z. Here S is the set of states, X is the set of inputs, Z is the set of outputs, $\nu(s, x)$ is the next state after a given state s and input x and $\delta(s, x)$ is the output from state s and input x. Its degree is sup {degree ν, degree δ}.

EXAMPLE 6.2.3.   Every finite state machine yields a finite state machine over 7 if 7 has 0, 1, where we take S, X, Z to be free modules and ν, δ transformation matrices.

EXAMPLE 6.2.4.   Let $7 = R^+ \cup \{e\}$ where e is an identity element and let $7 = S, X, Z$ and let $\nu(s, x) = s + x = \delta(x)$. Then we have a machine which can count or add.

Group choice theory is concerned with the problem of evaluating m alternatives by a group of n individuals. Let X be the set of alternatives. In the simplest case, each individual has a preference relation on X, expressing the pairs (x, y) such that he prefers x to y. This will be complete and transitive (a weak order). It will be a linear order if he values no two elements of X exactly the same. Then a group choi method gives a function from n-tuples of linear orders on X to some binary order on X. We write X as $\{x_1, x_2 \ldots , x_n\}$. Let P<i> be the matrix of person i's preference order.

DEFINITION 6.2.4.   A *social welfare function* with values in *7* is a function F from L^n to $M_m(7)$ where L is the set of m × m matrices of linear orders.

EXAMPLE 6.2.5.   If we let F be the matrix i, j such that $F_{ij} = 1$ if and only if $|\{k: P<k>_{ij} = 1\}| > \frac{n}{2}$ then we have majority voting.

We will assume that the main diagonal entries of (i, i) are identically 1. Let F_{ij} denote the (i, j)-entry of F.

DEFINITION 6.2.5.   That F is *independent of irrelevant alternatives* means if $P<k>_{ij} = R<k>_{ij}$ and $P<k>_{ji} = R<k>_{ji}$ then $F_{ij}(P_1, \ldots , P_n) = F_{ij}(R_1, \ldots , R_n)$

EXAMPLE 6.2.6.   Majority rule has this property.

PROPOSITION 6.2.4.   *F is independent of irrelevant alternatives if and only if F_{ij} is a function of $R<k>_i$*
Proof.   For linear orders $R<k>_{ij}{}^c = R<k>_{ji}$ so knowledge of $R<k>_{ij}$ is equivalent to knowledge of $R<k>_{ij}$ and $R<k>_{ji}$. The definition is equivalent to saying that F is a partial function of $R<k>_{ij}$ and $R<k>_{ji}$. But on $R<k>_{ij}$ we have a function since the domain is all n-tuples of $\{0, 1\}$.   □

DEFINITION 6.2.6.   That F is *neutral* means for any
permutation matrix P, $F(PR<1>P^T, \ldots, PR<n>P^T) =$
$PF(R<1>, \ldots, R<n>)P^T$. That F is *anonymous* means for
any permutation π, $F(R<\pi(1)>, \ldots, R<\pi(n)>) = F(R<1>,$
$\ldots, R<n>)$. More generally if this holds for a group
G of permutations we say F is G-*invariant*.

   EXAMPLE 6.2.7.   Majority voting has both properties.

   EXAMPLE 6.2.8.   If F is a constant nonsymmetric
relation then it is anonymous but not neutral.

   Anonymity is symmetry in the persons, while neutra-
lity is symmetry in the alternatives.

   DEFINITION 6.2.7.   A social welfare function is
transitive if and only if $F^2 \leq F$. It is *Pareto* if and
only if whenever $R<1> = \ldots = R<n>$, $F = R<n>$.

   This is a weak form of the Pareto property saying
that if all individuals have identical preferences,
the group preference must be the same as each individual's preference.

   EXAMPLE 6.2.9.   Majority rule has the Pareto property. Its transitive closure ΣF^n is transitive (but
no longer independent of irrelevant alternatives).

   EXAMPLE 6.2.10.   A constant function cannot have
the Pareto property.

   PROPOSITION 6.2.5.   *If F is independent of irrele-*
vant alternatives then it is also Pareto if and only
if $F_{ij}(0, \ldots, 0) = 0$ *and* $F_{ij}(1, \ldots, 1) = 1$.

Proof. The Pareto condition in this case is equivalent to saying that if all $R^{<k>}_{ij} = 0$ and all $R^{<k>}_{ji} = 1$ then $F_{ij} = 0$ and $F_{ji} = 1$. This proves sufficiency. But $R^{<k>}_{ij} = 0$ then $R^{<k>}_{ji} = 1$ so $F_{ij} = 0$. And if $R^{<k>}_{ij} = 1$ then $R^{<k>}_{ji} = 0$ so $F_{ij} = 1$. $\square$

THEOREM 6.2.6. *Social welfare function to 7 which are transitive, Pareto, independent of irrelevant alternatives, are in 1-1 correspondence with functions $f: V_n \rightarrow 7$ such that (i) $f(0) = 0$, (ii) $f(1) = 1$, (iii) if $v \le w$ then $f(v) \le f(w)$, and (iv) $f(vw) \ge f(v)f(w)$, where $F_{ij}(R^{<1>}_{ij}, \ldots, R^{<n>}_{ij}) = f(R^{<1>}_{ij}, \ldots, R^{<n>}_{ij})$ if there are at least three alternatives.*

Proof. Necessity of (i), (ii) for each function F_{ij} have been shown. Let u, v, w be any Boolean vector such that $uv \le w \le u + v$. Then we can construct a set of linear orders on any of these alternatives i, j, k, such that for all s, $R^{<s>}_{ij} = u_s$, $R^{<s>}_{jk} = v_s$, $R^{<s>}_{ik} = w_s$. To do this choose an ordering according to this table:

u_s	v_s	w_s	Order
0	0	0	kji
1	0	0	kij
0	1	0	jki
1	0	1	ikj
0	1	1	jik
1	1	1	ijk

Conversely, the conditions $uv \le w \le u + v$ are necessary since these are the only possible orders on i, j, k. So transitivity is equivalent to this equation $F_{ij}(u)F_{jk}(v) \le F_{ik}(w)$ whenever $uv \le w \le u + v$. This implies (iv) immediately. Let $u = 1$, $v = w$.

Then $F_{jk}(u) \leq F_{ik}(v)$. By symmetry F_{jk} and F_{ik} are
equal. Let $u = w$, $v = 1$. Then $F_{ij}(v) \leq F_{ik}(w)$. By
symmetry they are equal. This implies for $m > 2$ that
all F_{ij} equal to some function f. Moreover, if we let
$u = 1$ we obtain (iii). This proves necessity. Sufficiency follows also from these arguments. □

Many authors have obtained the following corollary.

DEFINITION 6.2.8.   A social welfare function is
oligarchical (*dictatorial*) if and only if there exists
a set S (a single person) such that for R<1>, ... ,
R<n>, F_{ij}(R<1>, ... , R<n>) = 1 if x_i R<k> x_j for all
k ε S and F_{ij} = 0 otherwise.

EXAMPLE 6.2.11.   If F(R<1>, ... , R<n>) = R<n> then
we have a dictatorial social welfare function.

COROLLARY 6.2.7.   *For* 7 = β, *all social welfare*
functions satisfying the conditions of the theorem are
oligarchical.

COROLLARY 6.2.8.   (Arrow's theorem.) *For* 7 = β, *a*
social welfare function is complete and satisfies these
conditions if and only if it is dictatorial.

In an oligarchical social welfare function, the
group prefers x_i to x_j if and only if every member of
the oligarchy does. Members outside the oligarchy have
no effect on group decisions.

PROPOSITION 6.2.9.   *Social welfare functions into*
7 *which are anonymous, independent of irrelevant*
alternatives, Pareto, and transitive are in 1-1

correspondence with functions $g: \{0\} \cup \underline{n} \to 7$ *such that* $g(0) = 0$, $g(1) = 1$, $g(n - a - b) \geq g(n - a)g(n - b)$ *and if* $a \leq b$ *then* $g(a) \leq g(b)$.

    Proof.   That a social welfare function is anonymous means f(v) depends only on the number of ones in v. Let g(a) be f(v) where v has a ones. Then the conditions of the last theorem translate directly to those here.   □

    PROPOSITION 6.2.10.   *Suppose* 7 *has no nilpotent elements. Then every Pareto, anonymous, and transitive social welfare function is a multiple of the unanimity function in which the group prefers i to j if and only if all individuals do.*

    Proof.   Exercise.

    Therefore, this holds for 7_1 and 7_3. However for 7_2 there does exist a social welfare function which precisely represents majority rule. Here 0, 1 are switched.

    EXAMPLE 6.2.12.   For 7_2 let F_{ij} be $\dfrac{k}{n}$ where k is the number of voters who do not prefer i to j.

    This gives a social welfare function which is transitive, Pareto optimal, neutral, and independent of irrelevant alternatives.

    Inclines can be used in decision theory. Let A = (a_{ij}) be the matrix of a decision table, that is a_{ij} is the value of choice i under state j of nature. Then a decision rule is to maximize $f(a_{i1}, \ldots , a_{in})$ where f is some function. For the maximin rule f is infimum. Then we find $\Sigma \ \Pi \ a_{i1} \ldots a_{in}$ for the incline 7_1 (or its dual). For the Nash bargaining solution (assigning 0 as disagreement value) we use the same formula for the incline 7_3. This maximizes the ordinary product $a_1 a_2 \ldots a_n$. Other inclines give other choice rules.

EXAMPLE 6.2.13.   We can apply this to a choice between two alternatives a, b under three possibilities 1, 2, 3:

$$
\begin{array}{cccc}
a & 0.4 & 0.7 & 0.8 \\
b & 0.5 & 0.5 & 0.5
\end{array}
$$

Then under the 7_1 choice rule we have sup $\{0.4, 0.5\}$ = 0.5 so b is chosen. But under the 7_3 (Nash) rule we have sup $\{(0.4)(0.7)(0.8), (0.5)^3\}$ = $(0.4)(0.7)(0.8)$ so a is chosen.

These rules can be applied to selection of an individual for promotion.

EXAMPLE 6.2.14.   Consider an individual's score on three attributes: human relationships, job performance, education. Let these be 0.2, 0.3, 0.8 for individual a; 0.3, 0.3, 0.7 for individual b; 0.6, 0.6, 0.6 for individual c. Then by the 7_3 choice rule we consider the supremum of the products 0.048, 0.063, 0.216 and so the last individual is chosen.

Ma and Cao (1982a, 1982b, 1983) give applications of inclines to multistage evaluation in psychological measurement and decision-making.

Logic is an algebra with operators and, or, if then, not, if and only if. The operations and, or are structurally analogous to sum, product in inclines. What about if then ? We can consider logic as a structure solely in this operation.

PROPOSITION 6.2.11. *A logical formula can be expressed solely in terms of $\to$ if and only if it has the form $p_1 \vee z$ for some basic variable z and formula p_1.*

Proof. From $y \to z$ we have $\sim y \vee z$ for all z. If $p_1 \vee z$ is obtained then $(p_1 \vee z) \to z$ gives $(\sim p_1 \wedge \sim z) \vee z = \sim p_1 \vee z$. If $p_1 \vee z$ and $p_2 \vee z$ are obtained we obtain $(p_1 \vee z) \to (p_2 \vee z)$ or $(\sim p_1 \wedge \sim z) \vee p_2 \vee z = \sim p_1 \vee p_2 \vee z$. So the class of p_1 which can be obtained contains $\sim y$ for all variables and is closed under negation and or. So it consists of all formulas p_1. If we have a formula $p_1 \to p_2$ then we have $\sim p_1 \vee p_2$. By induction we may suppose p_2 has the form $p_3 \vee z$. So all obtainable formulas have this form.   □

The operation $\vee$ can be defined solely in terms $\to$ by $x \vee y = (x \to y) \to y$ Therefore, if $\to$ is defined in a structure, it must be a semilattice under $\vee$.

PROPOSITION 6.2.12. *The element $z = x \to y$ in any Boolean algebra is uniquely characterized by (i) $z \geq y$, (ii) $z \vee x = 1$, and (iii) $z \wedge (x \vee y) = y$.*

Proof. Exercise.

These equations can define operations with some of the properties of $\to$ (if then). However, it does not exist in most inclines. For example, if the incline is linearly ordered, and $0 < x < 1$ and $x \not\leq y$, then (ii) implies $z = 1$ but this contradicts (iii) of Proposition 6.2.12.

6.3 APPLICATIONS TO GRAPHS, CLUSTERING, PROGRAMMING

Harary, Dorwin, and Cartwright introduced a method of calculating distances on graphs using matrices with a incline operation.

For a graph G, let $A = (a_{ij})$ be a matrix with real number entries such that a_{ij} is the directed distance along a single edge from i to j. Take $a_{ii} = 0$ and $a_{ij} = *$ if no such edge exists. We regard * as being infinity (or any very large number). Take this as a matrix over the incline $R^+ \cup \{0\}$.

PROPOSITION 6.3.1. *The (i, j)-entry of A^n gives the least distance from i to j along a path of length at most n.*

Proof. The (i, j)-entry of A^n is by induction the infimum of $a_{ii_1} + \ldots + a_{i_{n-1}j}$. But this is the distance from i to j along the edge sequence ii_1, i_1i_2, $\ldots$, $i_{n-1}j$. □

EXAMPLE 6.3.1. Let A over R^+, inf $\{x, y\}$, $x + y$ be

$$\begin{bmatrix} 0 & 0.4 & 0.3 \\ 0.1 & 0 & 0.5 \\ 0.6 & 0.7 & 0 \end{bmatrix}$$

Then A^2 is

$$\begin{bmatrix} 0 & 0.4 & 0.3 \\ 0.1 & 0 & 0.4 \\ 0.6 & 0.7 & 0 \end{bmatrix}$$

The distance from 2 to 3 by a length 2 path is a_{23}.

By tracing the products backward the *actual path* (*geodesics*) can be found.

The same method can be used in simple dynamic programming. Suppose we represent a process by a tree and that there is a cost assigned to each edge. Suppose that the total cost is the sum or product of the edge costs, or more generally is obtained by an incline product:

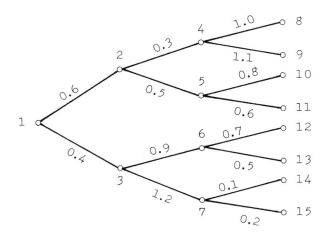

The product is to compute the least cost path to an
endpoint. In dynamic programming one first computes
the least cost paths from the vertices at level 2 to
the ends. Then one computes the least cost paths from
the level 1 vertices to the ends, and finally from the
level 0 vertex.

This problem can be dealt with also by matrices
using the same method as above. For each edge from i
to j enter the cost of that edge as a_{ij}. All other
edges have value * = ∞.

	1	2	3	4	5	6	7	8	9	10	11	12	13	14	15
1	*	.6	.4	*	*										
2	*	*	*	.3	.5										
3	*	*	*	*	*	.9	1.2	0	0						
4	*	*	*	*	*	*	*	1	1.1						
5	*	*	*	*	*	*	*	*	*	.8	.6				
6	*	*	*	*	*	*	*	*	*	*	*	.7	.5		
7	*	*	*	*	*	*	*	*	*	*	*	*	*	.1	.2

All other entries are *. This matrix can be operated
on as before. Take its square in the incline $R^+ \cup \{0\}$.

	1	2	3	4	5	6	7	8	9	10	11	12	13	14	15
1	*	*	*	.9	1.1	1.3	1.6								
2	*	*	*	*	*	*	*	1.3	1.4	1.3	1.1	*			
3	*	*	*	*	*	*	*	*	*	*	*	1.6	1.4		1.3 1.4

The desired distance will be determined by the one entry
of A^3v where $v = (*\ *\ *\ *\ *\ *\ *\ 0\ 0\ 0\ 0\ 0$
$0\ 0)^T$. The cube of A has as its only nonzero row
$(*\ *\ *\ *\ *\ *\ *\ 1.9\ 2\ 1.9\ 1.7\ 2\ 1.8\ 1.7$
$1.8)$. A^3v is then 1.7. There exist two paths yielding
this value, 1 to 11 and 1 to 14 since it is given by
the 11 and 14 entries of this row.

The dynamic programming solution applies also to all
games of perfect information. These can be represented
as a tree diagram and the best move found at any point.

Problems of control theory can be stated in a similar
way. Suppose we have a problem of continuous control
with discrete time intervals Δt. Let $u(t)$ be the control and $x(t)$ the controlled phenomenon, where $\Delta x =$
$g(x, u, t)\Delta t$ describes the path x takes over time.
Suppose it is desired to optimize a sum (approximating
an integral)

$$\sum_{n=0}^{s} F(x, u, n\Delta t)\Delta t, \quad s = (\Delta t)^{-1}$$

Let $c(y, t)$ be the optimum of

$$\sum_{n=\frac{t}{\Delta t}}^{s} F(x, u, n\Delta t)\Delta t, \quad s = (\Delta t)^{-1}$$

where x starts from $x(t) = y$. Then by dynamic program-
ming $c(y, t) = \min_{u} \{F(x, u, t)\Delta t + c(y + g\Delta t, t + \Delta t)\}$.
We can consider this as an operation over the incline.

If we let $\Delta t \to 0$ equations of optimality become $\frac{\partial F}{\partial u} + \frac{\partial c}{\partial y}$ $\frac{\partial g}{\partial u} = 0$ from the u derivative being zero and for the given path u(t), $0 = F + \frac{\partial c}{\partial y} g + \frac{\partial c}{\partial t}$.

The relation with Pontryagin's maximal principle is that the Hamiltonian $H(x, t, u) = F + \lambda(t)g$ where the Lagrange multiplier $\lambda(t) = \frac{\partial c}{\partial y}$ regarded as a function of t only. Then $\frac{\partial H}{\partial u} = 0$ is precisely the first equation above, and $\dot{\lambda} = - \frac{\partial H}{\partial x}$ is the x derivative of the second equation (with y taken as x); $0 = \frac{\partial}{\partial x} (F + \frac{\partial c}{\partial x} g + \frac{\partial c}{\partial t})$, $0 = \frac{\partial H}{\partial x} + g \frac{\partial^2 c}{\partial x^2} + \frac{\partial^2 c}{\partial x \partial t} = \frac{\partial H}{\partial x} + \frac{d}{dt} \frac{\partial c}{\partial x}$.

Y. Qin (1981a, 1981b, 1981c) introduced the *jarmetric principle* of which Bellman's optimality principle in dynamic programming, the Chapman-Kolmogorov equation for Markov decision chains, and the sub-optimal principle in multistage decision problems are special cases.

Maximin values can also be used to evaluate security levels of players and to evaluate the Nash bargaining solution.

Clustering is the problem of arranging a set of n objects labelled as n in disjoint subsets given as a matrix (d_{ij}) where d_{ij} for all i, j is a number describing the similarity of item i to item j. Ideally d_{ij} should be a symmetric matrix. A clustering being a partition of n, really corresponds to an equivalence relation. Clustering has been applied in medecine, biological classification, archaeology, and other areas. Its use is somewhat controversial. Perhaps it is best accepted as a tool in the sociology of small groups. It has been used to study scientists in the same discipline, members of a German town, and workers in a factory.

A family of clusterings can be obtained from any symmetric matrix A whose largest entries in each row lie on the main diagonal as follows. Choose any α and define a Boolean matrix B by $b_{ij} = 1$ if $a_{ij} \geq \alpha$ and $b_{ij} = 0$ if $a_{ij} < \alpha$. Then B will be reflexive and symmetric. And B^{n-1} will be idempotent, so it is the matrix of an equivalence relation.

However an equivalent result can be obtained in a simpler way.

DEFINITION 6.3.1.   A function $d(x, y)$ is an *-ultra-metric if and only if for all x, y, z: (i) $d(x, y) = d(y, x)$; (ii) $d(x, z) \geq \inf \{d(x, y), d(y, z)\}$, and (iii) $d(x, x) = d(y, y) \geq d(x, y)$ for all x, y.

A *-ultrametric is dual to what is called an ultrametric in the literature, that is, it is in terms of similarity rather than dissimilarity.

EXAMPLE 6.3.2.   This matrix is a *-ultrametric if we take $a_{ij} = d_{ij}$:

$$\begin{bmatrix} 2 & 1 & 0 \\ 1 & 2 & 0 \\ 0 & 0 & 2 \end{bmatrix}$$

DEFINITION 6.3.2.   A *hierarchical clustering* is a family of equivalence relations linearly ordered by inclusion.

EXAMPLE 6.3.3.   In biology, species, genus, family, etc. gives a hierarchical clustering.

PROPOSITION 6.3.2.   For a *-ultrametric the relations $E_\alpha = \{(x, y): d(x, y) \geq \alpha\}$ form a hierarchical clustering, where $\alpha \leq d(x, x)$.

Proof.   It is evident that they are linearly ordered. Condition (i) implies symmetry, condition (ii) implies

transitivity, and condition (iii) implies reflexivity.
☐

It can also be shown that every hierarchical cluster
ing arises from a *-ultrametric.

PROPOSITION 6.3.3.   *For any n-square symmetric matrix A, (I + A)$^{n-1}$ is a *-ultrametric.*

Proof.   Exercise.

EXAMPLE 6.3.4.

$$\begin{bmatrix} 1 & 0.5 & 0 & 0 \\ 0.5 & 1 & 0.3 & 0 \\ 0 & 0.3 & 1 & 0.5 \\ 0 & & 0.5 & 1 \end{bmatrix}^3 = \begin{bmatrix} 1 & 0.5 & 0.3 & 0.3 \\ 0.5 & 1 & 0.3 & 0.3 \\ 0.3 & 0.3 & 1 & 0.5 \\ 0.3 & 0.3 & 0.5 & 1 \end{bmatrix}$$

The hierarchical clustering consists of $\{1, 2\}$, $\{3, 4\}$
together with I, J.

Sometimes it is convenient to extend the fuzzy algebra so as to allow numbers larger than 1. Thus, all
we need to obtain is a reflexive, symmetric, and transitive Boolean matrix. One class of methods can be
described as follows: *given A, take its nth power A^n
according to some algebraic structure for some small n.
Then take the n - 1 fuzzy power of that.*
    The simplest such method is the *single-link* cluster-
ing method. In this we take A itself and derive B directly from it. Although this is a mothod in actual
use, it does have a disadvantage called *chaining*. Tha
is if two distinct clusters are connected by a single
chain of elements, the single-link method cannot dis-
tinguish them.

This can to some degree be avoided by taking the nth power first. However, if we take A^n as a power over the real numbers then another problem may occur: that large clusters obliterate smaller ones since they correspond to larger eigenvalues. Instead we suggest that A^2 be taken according to the structure $\{R^+, x + y, \inf \{x, y\}\}$ which is like an incline except that the operations are reversed. This method is somewhat similar to the method of taking k-connected components.

EXAMPLE 6.3.5.   This matrix consists of two clusters (and a separate point):

$$
\begin{bmatrix}
1 & 1 & 0 & 1 & 1 & 0 & 0 & 0 \\
1 & 1 & 1 & 0 & 1 & 0 & 0 & 0 \\
0 & 1 & 1 & 1 & 0 & 0 & 0 & 0 \\
1 & 0 & 1 & 1 & 1 & 0 & 0 & 0 \\
1 & 1 & 0 & 1 & 1 & 1 & 0 & 0 \\
0 & 0 & 0 & 0 & 1 & 1 & 1 & 1 \\
0 & 0 & 0 & 0 & 0 & 1 & 1 & 1 \\
0 & 0 & 0 & 0 & 0 & 1 & 1 & 1
\end{bmatrix}
$$

Its square under the given structure is

$$
\begin{bmatrix}
4 & 3 & 2 & 3 & 3 & 1 & 0 & 0 \\
3 & 4 & 2 & 3 & 3 & 1 & 0 & 0 \\
2 & 2 & 3 & 2 & 2 & 0 & 0 & 0 \\
3 & 3 & 2 & 4 & 3 & 1 & 0 & 0 \\
3 & 3 & 2 & 3 & 5 & 2 & 0 & 1 \\
1 & 1 & 0 & 1 & 2 & 4 & 3 & 3 \\
0 & 0 & 0 & 0 & 0 & 3 & 3 & 3 \\
0 & 0 & 0 & 0 & 1 & 3 & 3 & 3
\end{bmatrix}
$$

Then iterating this over the fuzzy algebra (which we extend by allowing values greater than 1), and

increasing main diagonal entries to 5 to make it reflex
ive, we have

$$
\begin{bmatrix}
5 & 3 & 2 & 3 & 3 & 2 & 2 & 2 \\
3 & 5 & 2 & 3 & 3 & 2 & 2 & 2 \\
2 & 2 & 5 & 2 & 2 & 2 & 2 & 2 \\
3 & 3 & 2 & 5 & 3 & 2 & 2 & 2 \\
3 & 3 & 2 & 3 & 5 & 2 & 2 & 2 \\
2 & 2 & 2 & 2 & 2 & 5 & 3 & 3 \\
2 & 2 & 2 & 2 & 2 & 3 & 5 & 3 \\
2 & 2 & 2 & 2 & 2 & 3 & 3 & 5
\end{bmatrix}
$$

This hierarchical clustering gives I, J, and at $\alpha = 3$,
$\{1, 2, 4, 5\}$, $\{3\}$, $\{6, 7, 8\}$.

In place of the matrix A^2, one can take the correlation matrix of A with itself, an iterated correlation
matrix or just divide all $a_{ij}^{(2)}$ by $\sqrt{a_{ii}a_{jj}}$ and take a
higher power. This is a little like the CONCOR algo-
rithm which iterates correlation matrices of the rows of A.

A problem closely related to clustering theory is
that of assigning an additional object to one of n
existing clusters. This occurs in pattern recognition
when a new pattern must be classified by its closeness
to existing clusters.

In principle this can be solved by clustering the
group with the new object added.

PROPOSITION 6.3.4.   *Suppose objects are grouped into
n clusters at level α by the single-link method. Suppose another object k is to be added to the clusters
at some level $\gamma \geq \alpha$ at which the clusters are still
distinct by clustering the total group. Then that
object must be added to the cluster which has an objec
i most similar to n + 1 (i.e., $d_{ik} = \max_{j} \{d_{jk}\}$).*

Proof. Suppose another cluster has an object j so that $d_{jk} \leq d_{ik}$. Then by the single-link method if i is to be identified with k, so must j. Therefore the two clusters coincide. But this is contrary to assumption. $\square$

EXAMPLE 6.3.6. Suppose 3 objects are given with this data matrix, clustered as {1}, {2, 3}:

$$\begin{bmatrix} 1 & 0.2 & 0.3 \\ 0.2 & 1 & 0.5 \\ 0.3 & 0.5 & 1 \end{bmatrix}$$

Suppose an additional object has distances (0.4, 0.45, 0.42) from the 3 objects. Then it must be clustered with {2, 3}, since it is most similar to {2}.

This idea can extend to the more general methods in which one performs an operation on A, and then applies single-link clustering. Then an object will go to the cluster having an element which maximizes the entry of the resulting matrix. So if we take the matrix power according to some algebraic structure we must maximize $A_{k*}A_{*i}$. If we take the correlation matrix we must maximize the correlation of A_{k*} and A_{*i}.

6.4 ANTIINCLINES AND APPLICATIONS TO MATRIX THEORY

Inclines can be used in dealing with matrices in applications of matrix theory. These applications are analogous to those of Boolean matrices to nonnegative matrix theory. In the Boolean case to a matrix A we assign (b_{ij}) where $b_{ij} = 0$ if $a_{ij} = 0$ and $b_{ij} = 1$ if $a_{ij} = 1$. This mapping is a semiring homomorphism, so that B^n describes precisely which entries of A^n are nonzero.

There exists a concept dual to inclines called *antiincline*. Antiinclines differ from inclines only in that we require $xy \geq x$ and $xy \geq y$. These are semilattice-ordered commutative semigroups in which the product is greater than or equal to the factors. However simply reversing the order will not in general change an antiincline into an incline.

Distributivity $x(y \wedge z) = xy \wedge xz$ is neither necessary nor sufficient in general for linear representability.

EXAMPLE 6.4.1. The law holds for the incline with basis 1, x, y, x^2, xy, y^2 and the relation $(x + y)^3 = 0$ yet $xy \leq x^2 + y^2$ is not true. The only incomparable w, z such that $w \wedge z > 0$ are x, y.

EXAMPLE 6.4.2. Let $7 = 7_1 \times 7_1$ where $7_1 = \{1, x, x^2, x^3, x^4, x^5, x^6 = 0\}$. Let 7_2 be the subincline generated by $\{(x^2, x^2), (x^2, x^3), (x^3, x^2), (x^5, x^5)\}$. Then 7_2 is closed under sums and products. We have $(x^2, x^2)\big((x^2, x^3) \wedge (x^3, x^2)\big) = (x^2, x^2)(x^4, x^4) = 0 < (x^4, x^5) \wedge (x^5, x^4) = (x^5, x^5)$.

Antiincline structures occur in a number of situations in mathematics (where a few exceptions are allowed).

EXAMPLE 6.4.3. Let $v_p(x)$ for a rational number x denote the power of the prime p dividing x. Then $v_p(xy) = v_p(x) + v_p(y)$ and $v_p(x + y) \geq \inf \{v_p(x), v_p(y)\}$. And in the general case, equality holds. With equality we have an incline structure.

EXAMPLE 6.4.4.   Consider the degree of a polynomial
or rational function.   Then degree (fg) = degree (f) +
degree (g) and degree $(f + g) \leq$ sup {degree (f), degree
(g)} with equality holding in the general case.   With
equality this gives an antiincline structure.

Incline theory then can yield results about divisi-
bility and degrees of polynomials.

EXAMPLE 6.4.5.   Consider the nth power of

$$\begin{bmatrix} x^2 + 1 & x \\ x & x^2 + 3 \end{bmatrix}$$

The degrees of the various entries will equal those in
the nth power of

$$\begin{bmatrix} 2 & 1 \\ 1 & 2 \end{bmatrix}$$

over the antiincline structure on R^+.

Here equality certainly holds because no terms are
negative.

EXAMPLE 6.4.6.   The $(1, 2)$-entry of

$$\begin{bmatrix} 4 & 2 \\ 3 & 6 \end{bmatrix}^5$$

will be divisible by 2^3 since 2 is the $(1, 2)$-entry of

$$\begin{bmatrix} 2 & 1 \\ 0 & 1 \end{bmatrix}^5 = \begin{bmatrix} 3 & 3 \\ 2 & 3 \end{bmatrix}$$

over the incline.

Another application of incline theory to Boolean
matrix theory involves matrices partitioned into $r \times r$

blocks. Suppose each block belongs to a subsemiring of B_r, which is an incline (antiincline). Here B_r denotes the set of all r-square Boolean matrices. Then we can regard the matrix as a matrix over this incline of smaller dimension.

EXAMPLE 6.4.7.   This matrix

$$\begin{bmatrix} 1 & 0 & 0 & 0 \\ 0 & 1 & 0 & 1 \\ 0 & 0 & 0 & 0 \\ 0 & 1 & 0 & 0 \end{bmatrix}$$

can be regarded as a matrix over fuzzy algebra:

$$\begin{bmatrix} 1 & 0.5 \\ 0.5 & 0 \end{bmatrix}$$

The set of all reflexive Boolean matrices forms an antiincline since $AB \geq AI$ and $AB \geq IB$. In fact any commuting set of n-square reflexive Boolean matrices forms an antiincline.

Inclines also occur as sets of ideals.

THEOREM 6.4.1.   *Suppose all two-sided ideals in a semigroup commute. Then they form an incline under the operations $M \cup N$, MN where M and N are ideals. If the two-sided ideals in a semiring with o commute then they form an incline under the operations $M + N$, MN.*

Proof.   The commutative, associative, and distributive properties are standard. We have $M \cup M = M$ and $M + M = M$ since $0 \in M$ and M is closed under addition. Since $MN \subset M$, $M + MN = M$.   □

EXAMPLE 6.4.8.   This is not true for all Lie algebras. One can form a Lie algebra which is nilpotent of class 2 in which associativity of ideals is not true.

Permanents are related to the assignment problem of linear programming. We consider expected values and maximum and minimum values of permanents.

DEFINITION 6.4.1.   Let S be an arbitrary set. Let $A \in M_n(S)$. Then for π ranging over all permutations,
$$\text{per}_S (A) = \sum_\pi \prod_1^n a_{i\pi(i)}.$$

EXAMPLE 6.4.9.   For $A \in M_2(S)$, per $(A) = a_{11}a_{22} + a_{12}a_{21}$.

The assignment problem is as follows: *assign n people to n tasks.* Let a_{ij} be the productivity of person i in task j. Suppose the joint productivity is measured by a function such as the product over 7. Then maximize joint productivity.

The solution to the assignment problem results in productivity precisely $\text{per}_7 (A)$.

DEFINITION 6.4.2.   A *doubly stochastic matrix* is one whose row and column sums over the real numbers are all 1.

THEOREM 6.4.2.   *The minimum value of* $\text{per}_7 (A)$ *over doubly stochastic matrices is for* 7_3 n^{-n}*, for* 7_2 0 *for* 7_1 $\left((1 + [\frac{n}{2}]) (n - [\frac{n}{2}]) \right)^{-1}$.

For a random matrix over 7, the distribution of the permanent and similar quantities is unknown.

6.5 OPEN PROBLEMS

1. How can we consistently estimate coefficients in a linear equation over an incline by a simple

algorithm ? That is over the real numbers if y
$= a_1x_1 + \ldots + a_nx_n + u$ where u is a random error
$E(u) = 0$ then we can find a_i approximately by
least squares from observations of y, $x_1, \ldots$, x
However, the relation y = inf $\{a_ix_i\}$ is more complicated. If we require that for the actual obse
vations y $\geq$ inf $\{a_ix_i\}$ such as results for a gene
alized inverse, then the estimate cannot be cons
tent, i.e., it will not generally converge to the
true values of the a_i. The same is true if we tr
to minimize sup $|y - \inf \{a_ix_i\}|$ or sup $|\frac{y}{a_ix_i}|$.
If we take least squares, the problem is that
there is no simple way to solve the resulting
equations. Suppose we minimize Σ $|\log y - \inf$
$\{\log a_i + \log x_i\}|$. Can this be solved by linear
programming ?

2. How can we determine the number of variables in
such a relation and whether one exists ?

3. What algorithms can efficiently be performed by
a finite machine over **7** ?

4. A determinant mapping would be a homomorphism
$M_n(7) \to 7$ under multiplication. One is given by
h(x) = 1 if x is invertible and h(x) = 0 otherwise. Are there others ?

5. Develop a method of estimating the permanent of
a matrix of real numbers. We can use row sums,
solutions to various flow problems, inclusion-
exclusion formulae. There are Monte Carlo method
attach random signs and compute the determinant,
take random permutations π to estimate the avera
value of Π $a_{i\pi(i)}$.

References

D. Allen, Jr., A generalization of the Rees theorem
to a class of regular semigroups, Semigroup Forum 2
(1971), 321-331.

G. Birkhoff, Lattice Theory, Amer. Math. Soc., Colloq.
Pub. Vol. 25, Providence, R. I., 1967.

Zhi-qiang Cao, An algebraic system generating the
fuzzy subsets of a set, in Advances in Fuzzy Set Theory
and Applications, P. P. Wang, Ed., to appear.

Zhi-qiang Cao, Modern control theory of psychological
phenomena, Proceedings of CAA-IEEE Bilateral Meeting
on Control Systems, August 10-12, 1981, Shanghai.

Zhi-qiang Cao, Modern control theory of psychological
phenomena, in Recent Developments in Control System
and its Applications, Chinese Association of Automation,
Ed., Gordon and Breach, New York, 1982, 182-190.

Zhi-qiang Cao, Introduction to modern control theory
of psychological phenomena, in A General Survey of
System Methodology, L. Troncale, Ed., Society for
General Systems Research, Washington, 1982, 1042-1045.

Zhi-qiang Cao, Several results on ideal of algebraic
incline, Science Monthly (China), 28/14 (1983), 8920.

160   References

B. Carre, Graphs and Networks, Clarendon Press, Oxford
1979.

A. Clifford and G. Preston, The Algebraic Theory of
Semigroups, Amer. Math. Soc., Providence, R. I., 1961

R. Cuninghame-Green, Minimax Algebra, Lecture Notes,
No. 166, Springer-Berlag, Berlin, 1979.

D. Dubois and H. Prade, Fuzzy Sets and Systems Theory
and Applications, Academic Press, New York, 1980.

L. Fuchs, Partially Ordered Algebraic Systems, Pergamo
Oxford, 1963.

H. Hamacher, Maximal algebraic flows: algorithms and
examples, in Discrete Structures and Algorithms, U.
Pope, Ed., Hanser, Munich, 1980.

H. Hamacher, Algebraic flows in regular matroids,
Discrete Applied Mathematics, 1(1980), 27-38.

H. Hashimoto, Symmetric kernels and nilpotent parts of
Boolean matrices representing transitive relations,
Preprint, Yamaguchi University, Yamaguchi, Japan, 1980

H. Hashimoto, Idempotent Boolean matrices and transitive reduction, Preprint, Yamaguchi University,
Yamaguchi, Japan, 1981.

H. Hashimoto, Transitivity of matrices over a path
algebra, Preprint, Yamaguchi University, Yamaguchi,
Japan, 1983.

H. Hashimoto, Reduction of a fuzzy retrieval model,
Information Sciences, 27(1982), 133-140.

H. Hashimoto, Reduction of a nilpotent fuzzy matrix,
Information Sciences, 27(1982), 233-243.

H. Hashimoto, Decomposition of Boolean matrices and
its applications, Trans. of IECE (Japan), E66(1983),
39-46.

H. Hashimoto, Canonical form of a transitive fuzzy
matrix, Fuzzy Sets and Systems, 11(1983), 157-162.

H. Hashimoto, Subinverses of fuzzy matrices, Fuzzy
Sets and Systems, 12(1984), 155-168.

A. Kaufmann, Introduction to the Theory of Fuzzy Subsets, Academic Press, New York, 1965.

K. H. Kim, Boolean Matrix Theory and Applications, Marcel Dekker, New York, 1982.

K. H. Kim and F. W. Roush, Some results on decidability of shift equivalence, Jour. of Comb., Infor. & System Sci., 4(1978), 35-52.

K. H. Kim and F. W. Roush, Introduction to Mathematical Consensus Theory, Marcel Dekker, New York, 1980.

K. H. Kim and F. W. Roush, Generalized fuzzy matrices, Fuzzy Sets and Systems, 4(1980), 293-315.

K. H. Kim and F. W. Roush, Fuzzy flows on networks, Fuzzy Sets and Systems, 8(1982), 35-38.

K. Krohn and J. Rhodes, Complexity of finite semigroups, Annals of Math., 88(1968), 128-160.

K. Krohn, J. Rhodes, and B. Tilson, Lectures on the algebraic theory of semigroups and finite state machines, in Algebraic Theory of Machines, Languages, and Semigroups, M. A. Arbib, Ed., Academic press, New York, 1968.

C.-Z. Luo, Generalized inverses of fuzzy matrices, in Approximate Reasoning in Decision Analysis, M. M. Gupta and E. Sanchez, Eds., North-Holland, Amsterdam, 1982.

M.-C. Ma and Z.-Q. Cao, Multistage evaluation method in psychological measurements, in Approximate Reasoning in Decision Analysis, M. M. Gupta and E. Sanchez, Eds., North-Holland, 1982, 307-312.

M.-C. Ma and Z.-Q. Cao, A fuzzy model for category judgment and the multistage evaluation method, Acta Psychologica Sinica, 15(1983), 198-204 (Chinese).

M.-C. Ma and Z.-Q. Cao, A kind of multidimensional decision making and its mathematical model, in General Survey of Systems Methodology, L. Troncale, Ed., Society for General Systems Research, Washington, 1982, 1036-1041.

H. Mortazavian, On k-controllability and k-observability, Preprint, Univ. of Toronto, Canada, 1983.

References

C. V. Negoita and D. Ralescu, Applications of úzzy Se
to Systems Analysis, Wiley, New York, 1975.

Y. Qin, On jar metric principle (I), Science Exploration, 1(1981), 59-76.

Y. Qin, Jar metric principle (II), Science Exploratio,
1(1981), 101-108.

Y. Qin, Algorithms on optimal path problems, Science
Exploration, 1(1984), 61-94.

B. Tilson, Decomposition and complexity of finite
semigroups, Semigroup Forum, 3(1971), 189-250.

R. F. Williams, Classfication of subshifts of finite
type, Ann. Math., 2(1974), 380-381.

F. Ramsey, On a problem of formal logic, Proc. London
Math. Soc., 30(1930), 204-286.

J. Rhodes, Finite binary relations have no more compl
xity than finite functions, Semigroup Forum, 7(1974),
92-103.

J. Rhodes and B. Tilson, Lower bounds for complexity
of finite semigroups, Jour. of Pure and Appl. Alg.,
1(1971), 79-95.

R. R. Yager, On a general class of connectives, Fuzzy
Sets and Systems, 4(1980), 235-242.

L. A. Zadeh, K.-S. Fu, K. Tanaka, and M. Shimura, Eds
Fuzzy Sets and Their Applications to Cognitive and
Decision Processes, Academic Press, New York, 1975.

U. Zimmermann, Linear and Combinatorial Optimization
in Ordered Algebraic Structures, Annals of Discrete
Mathematics, No. 10, North-Holland, Amsterdam, 1981.

Index

My
FUTURE
CAREER

Working with
Computers

Margaret McAlpine

GARETH**STEVENS**
GS
PUBLISHING
A World Almanac Education Group Company

Please visit our web site at: **www.garethstevens.com**
For a free color catalog describing Gareth Stevens Publishing's
list of high-quality books and multimedia programs, call
1-800-542-2595 (USA) or 1-800-387-3178 (Canada).
Gareth Stevens Publishing's fax: (414) 332-3567.

Library of Congress Cataloging-in-Publication Data

McAlpine, Margaret.
    Working with computers / Margaret McAlpine.
      p. cm. — (My future career)
    Includes bibliographical references and index.
    ISBN 0-8368-4242-1 (lib. bdg.)
      1. Electronic data processing—Vocational guidance—Juvenile literature.
    2. Computer science—Vocational guidance—Juvenile literature. I. Title.
    QA76.25.M43   2004
    004'.023—dc22   2004045227

This edition first published in 2005 by
Gareth Stevens Publishing
A World Almanac Education Group Company
330 West Olive Street, Suite 100
Milwaukee, Wisconsin 53212 USA

This U.S. edition copyright © 2005 by Gareth Stevens, Inc. Original
edition copyright © 2004 by Hodder Wayland. First published in 2004
by Hodder Wayland, an imprint of Hodder Children's Books.

Editor: Dorothy L. Gibbs
Inside design: Peta Morey
Cover design: Melissa Valuch

Picture Credits
Corbis: Phil Banko 19(r); Brooklyn Productions 29; Corbis: 9, 10, 13, 15, 16, 21,
24, 37, 39, 51, 56; Jim Craigmyle 40; Darama 44; DiMaggio/Kalish 43(t); Paul
Edmondson 17; ER Productions 30; Randy Faris 19(l), 48; John Feingersh 22, 49;
Patrik Giardino 59(r); Tom Grill 35(t); Walter Hodges 23, 47; Ted Horowitz 55;
JFPI Studios, Inc. 35(b); Jose Luis Pelaez, Inc. 5, 6, 27(t), 28, 32, 53; Catherine
Karnow 59(l); Helen King 33; Andrew Kolesnikow, Elizabeth Whiting and Associates
43(b); Larry Williams and Associates 57; Lester Lefkowitz 38; Left Lane Productions
31; Rob Lewine 27(b); Robert Maass 45; Michael A. Keller Studios, Ltd. 52; Darren
Modricker 25; Paul Morris 50; Joaquin Palting 7, 41; Picture Arts 11, 20; Michael
Pole 54; Michael Prince 46; Anthony Redpath 12; Gerhard Steiner 4, 36; Bill Varie
14. **Corbis Saba:** Tom Wagner 8. **Getty Images:** cover.

Gareth Stevens Publishing thanks the following individuals and organizations
for their professional assistance: Chris Crawford, author of fourteen published
games and seven books on computing; Norstan Communications, Inc. (Lori
Moore, Information Systems Professional; Kevin Terveen, Senior Technical
Engineer/Cisco Support Team; Steven Fuchs, Senior Technology Manager;
Bill Horejs, Professional Services Manager; Christine Holloway, Director of
Marketing).

Printed in China

1 2 3 4 5 6 7 8 9 08 07 06 05 04

Contents

Words that appear in the text in **bold**
type are defined in the glossary.

Games Designer

What is a games designer?

Games designers are people who create and write the **software** for games that are played on computers or on popular game machines such as Game Boy, Xbox, and PlayStation. The **programming** for modern computer games is so advanced that games designers almost never work alone. Instead, they work as members of games design teams.

A large number of people, today, play computer games on a regular basis, and they have an incredible variety of games to choose from. So many different types of games are produced that it can truly be said there is something for everyone.

Not only are there lots of different games, but also, many games are available for use on a number of different computer **platforms** or types of game machines, including:

- PCs and Macs
- Game Boy and Game Boy Advance
- GameCube
- PlayStation and PlayStation 2
- Xbox

A brainstorming session is often the first step in developing a computer game.

That's Games-Biz!

The video and computer games industries are big business. At present, in the United States, more money is made from sales of computer games and related products than from movie box office sales, which seems to indicate that many people prefer video games to movies.

Making a computer game available for use on more than one platform means rewriting the game's programming for each different platform, even if each version of the game will be exactly the same. Sometimes, rewriting a computer game will also mean using a completely different **programming language** for each platform.

When a computer game sells well, it is not unusual for its designers to go back into the program to create an "expansion pack," adding extra levels to the game and, often, introducing new characters.

Artists draw their impressions of the characters for a new game before the game's software is written.

Main responsibilities of a games designer

Almost every new computer game starts with a brain-storming session in which everyone in a games design team suggests new ideas for games. The best ideas are developed into concepts, and the game concepts are shown to a panel of experienced staff members. If the panel likes a concept, that game moves on for further development.

An early stage of any game development project is making a prototype, or rough sample, of the game to show its features. The main features of a prototype include:

- the look, or appearance, of the game
- the game's main characters
- the movements of the characters
- the way in which the game is played

The development process is usually a team effort, and design plans are always subject to change.

Good Points and Bad Points

"Being a games designer gives me an opportunity to work with imaginative people, and taking ideas for new games and turning them into reality is exciting. It's also good seeing a positive review for a game I have worked on. Of course, it's not so good when a review isn't positive."

After a prototype is approved, the game's designers create all the detailed stages of the game, which include the following features:

- environment, or setting
- speech used by the characters
- **user interface**, which is the information for players that appears on the screen

To write the best video and computer games, designers must keep up to date with **innovations** and changes in computer **hardware**.

Before they begin to write the programming **code** for a game, designers break down the game's story line into stages. Each stage becomes a level of the game. Once the original code is written, designers supervise testing of the game. A game must be tested to make sure it runs well on the platform for which it was written. Testing usually results in any number of changes or additions to a game's software.

With all of the testing and changes, designing a game can take a long time. Games often spend months in the design phase and take at least a year to complete. The development time for big games is sometimes more than two years.

Main qualifications of a games designer

Imagination

Coming up with ideas and concepts for new games requires creativity and a vivid imagination.

Technical knowledge

To know what makes a video game or computer game successful, a games designer has to know all about the history of electronic games and how they were developed. Advanced knowledge of computers (both hardware and software) and programming languages is also important. A games designer's technical knowledge must always be up to date so the games he or she produces will be at the forefront of technology.

After the code is written, a game must be tested, again and again.

Analytical ability

Games designers need well-developed analytical skills. They have to deal with a lot of small details, and they must know how to locate problems as well as how to correct them.

Communication skills

Good communication within a game's design team is extremely important. The various members of the team are each responsible for different aspects of the game, such as character appearance, game controls, or sound. All parts of the game need to work together perfectly for the game to be a success, which means all members of the team have to effectively communicate to each other their ideas, what they are doing, and how they are progressing.

**A games designer's goal
is to produce hours of fun.**

Teamwork

Being a games designer is not
a job for anyone who wants
to work alone. Developing a
game is a long process and too
complex for just one person.

Commitment

Competition for jobs in
the games industry is huge.
Designers must be willing to
put in long hours for low pay.

fact file

Along with a creative mind and
lots of experience with computer
games, a designer should have a
college degree and lots of course
work in computer science. Even
then, the path to a career usually
means finding a low-level job
with a games company and
slowly working up to designing.

Dave Alvarez

Dave is twenty-nine years old. He has been a games enthusiast for many years and now works for a games development company.

7:30 a.m. I'm on the road, headed for the office. I have a long drive to work, and I often spend the time thinking over ideas for games.

8:30 a.m. I arrive at the office and read through my E-mails. If possible, I reply to them right away. Otherwise, I mark them for action later in the day.

9:00 a.m. I start getting ready for a development team meeting by entering testing results for a **PC-based** game into a **database**.

Testing for this game has been going on for some time. We just finished testing for possible **compatibility** problems. This type of problem occurs when software doesn't work well with certain hardware components.

11:00 a.m. Now I start creating the presentation materials for this afternoon's team meeting. The database into which I have been entering testing results automatically produces figures and turns them into attractive charts and graphs.

Some computer games become so popular that they are made into films.

A games designer's busy schedule can often mean working through lunch.

12:00 p.m. I spend my lunch hour responding to E-mails.

1:00 p.m. It's time for the development team meeting. I show team members the results of the PC game's compatibility tests. After discussing all of the problems, we decide which are the most urgent. Then we talk about what we can do to solve the problems and what effect coding changes might have on the game. We also decide who will be responsible for making changes to the game.

5:00 p.m. I'm at my computer. Guess who the team decided should make the coding changes?

7:00 p.m. I've turned off my computer but find myself sketching out some of the ideas I had on my way to work this morning.

Hardware Engineer

What is a hardware engineer?

Hardware engineers are the people responsible for setting up, maintaining, repairing, and updating the equipment that makes up computer systems. A computer system's hardware includes all of the system's various electronic parts, such as its central processing unit (CPU), **monitor**, keyboard, cables, **network server**, and **modem**.

Companies with large computer systems often have Information Technology (IT) departments that include teams of hardware engineers. Having in-house engineers means that problems can usually be taken care of soon after they arise. Hardware engineers working within a company are often scheduled on different **shifts** or must be on call so that an engineer is always available to take care of emergencies.

Many hardware engineers are self-employed and travel to wherever the clients who need their skills are located. Some hardware engineers travel to locations across the country or even abroad. Work at a client's site can last from a few hours to a month or longer.

The ongoing development of more complex and powerful **software** means a constant need for better and faster computer systems.

Being a hardware engineer requires a high level of technical skill and computer knowledge.

The Computer Evolution

In 1943, the chairman of the IBM Corporation, which, today, is one of the world's foremost computer manufacturers, believed there was worldwide market potential for "maybe five computers." By the 1960s, top engineers were saying that, in the future, it might be possible to produce computers weighing no more than one and a half tons.

Today, the world has countless computer systems, some so small and lightweight that they fit inside wristwatches and mobile phones.

When the original manufacturers no longer provide fixes and **upgrades** for older systems, hardware engineers must update them. Updating an older system can be challenging, however, because hardware engineers are often forced to meet the specific requirements of the companies that sell the equipment. When older hardware becomes too likely to break down, the system should be completely replaced. In these cases, hardware engineers will write reports that describe the problems and possible solutions, but it is the clients who make the decisions to repair or replace.

For a hardware engineer, no job is too big or too small. Even home computer systems develop problems that may require an engineer's skills.

Main responsibilities of a hardware engineer

Along with keeping computer equipment in good working order, a hardware engineer's responsibilities may include:

- regularly examining the hardware being used to **monitor** its performance and efficiency
- determining when upgrades and new components are required and calculating their costs
- finding the most effective ways to update a system
- deciding when an entire system needs to be replaced and providing recommendations and cost estimates for new hardware
- training clients to use new or upgraded hardware or software

Equipment breakdowns mean lost work time and cost companies a lot of money so hardware engineers must locate and resolve problems as quickly as possible.

Being on call can mean a long workday.

Good Points and Bad Points

"As a hardware engineer, I have to make sure that computer equipment is operating properly. The best part of my job is seeing all the different parts of a computer system working together smoothly. Often, however, I am faced with problems that are not easy to solve, and finding the answers can sometimes be very frustrating."

A hardware engineer will often have to work with financial **analysts**. Because the equipment hardware engineers maintain is so expensive, they must prove that the benefits of the work they do are worth the cost.

After recommended hardware upgrades or changes have been approved, engineers are responsible for their **installation**. This stage of the project may take some time, especially if a large network is being replaced.

A project isn't necessarily finished, either, after the new hardware has been installed. Engineers often have to deal with **compatibility** problems, which occur when system components do not work well together. Sometimes, these problems are not noticeable until after the hardware has been in use for a while so a hardware engineer may be needed to make adjustments even long after installation has been completed.

Hardware engineers have to constantly keep up with new technology.

One other important area of responsibility for hardware engineers is **virus** protection, and when systems become infected, engineers must determine a course of action to eliminate the virus and restore the system.

Main qualifications of a hardware engineer

Technical knowledge and mechanical skill

Maintaining complicated computer systems requires a great deal of technical knowledge and the ability to work with machines and tools. Also, because the world of computers is constantly changing, hardware engineers have to spend time keeping up with new developments in information technology.

Analytical ability

To figure out what is wrong with a computer system, hardware engineers need good analytical skills. When upgrading systems, they have to know the hardware requirements for the software applications that will be used and know how to choose precisely the right upgrades to make the system run efficiently.

Many hardware engineers work as part of a team. Members of the team must share information with each other on a regular basis.

Time management

Hardware engineers often have to meet deadlines set by their companies or clients. Because the deadlines can be very tight, engineers have to be committed to their work and prepared to put in long hours to complete a job on time.

Communication skills

Many of the people hardware engineers have to work with do not have a lot of technical knowledge so the engineers need strong communication skills to be able to explain technical information clearly. They also must be able to get along with all different kinds of people, including dissatisfied customers.

Adaptability

Work on a computer system can often change a great deal between planning and completing a job. For this reason, hardware engineers need to be flexible, and they have to be able to make changes easily and quickly.

High standards

Whether they are self-employed or working in-house, hardware engineers have to be organized, work quickly, and maintain high standards of quality.

When dealing with computer hardware problems, investigation is the first step in determining a plan of action.

fact file

Hardware engineers need highly specialized education and training, including at least a bachelor's degree in computer science, computer systems engineering, or electrical engineering. With so many students graduating in these fields, however, jobs are very competitive, so an advanced degree and work experience are strongly recommended.

Dennis Chung

Dennis is a hardware engineer working for a company that helps large corporations solve problems they might be having with their computer systems.

8:30 a.m.   Before going out on any calls, I spend some time checking phone and E-mail messages and keeping myself up to date on new hardware and software developments.

9:00 a.m.   A customer phones, telling me that all is well with some network server repairs I did yesterday. I am relieved to know that the customer is happy with my work. I respond to some E-mailed requests and return a few phone calls.

10:00 a.m.   I take a call from a customer who runs the IT department at an engineering factory. He is having a problem with his **floppy disk drive**. After discussing the problem, I am able to suggest a solution. Some difficulties can be handled over the phone, but I spend most of my time working on customers' **premises**.

10:30 a.m.   An important corporate customer calls to report a problem with the company's $2.5 million hardware. This problem is a high priority and needs my immediate attention.

We have different agreements with different customers regarding how we prioritize problems. The agreements vary from fixing a problem within a few hours to having a problem taken care of within twenty-four hours or more. Customers usually pay more for higher priority agreements.

11:00 a.m. I'm in my car and on my way, armed with my toolbox, which I take on every job.

11:30 a.m. I open up the computers and approach the repair in much the same way that an electrician would repair a refrigerator or a washing machine. I also have a set of CDs that I can put into the computer to help me **diagnose** what is wrong.

2:00 p.m. I'm making progress, but it's a long job. I also have to spend time reassuring the staff that I know what I'm doing.

4:15 p.m. Back in my office, I write a report on the problem I've just resolved. Then I check with my morning caller to find out if the solution I suggested has taken care of his floppy disk problem.

6:00 p.m. I close my office, taking a pager with me because, tonight, I'm on call for emergency breakdowns.

Hardware engineers must know all about many different kinds of computer components.

Server rooms like this one are a common sight today.

Help Desk Professional

What is a help desk professional?

Help desk professionals respond to questions from people who are having problems with their computer **hardware** or the **software** applications they are using. Usually, the help desk professional first listens to a description of the problem, then suggests ways to solve the problem. Most often, help is provided by giving step-by-step instructions over the phone.

There are several different types of help desk professionals:

- A company that designs and produces software applications needs help desk professionals to support users who are having difficulty **installing** or operating the company's software.
- A company that produces computer hardware needs help desk professionals to resolve problems for customers who are trying to add the hardware to their computer systems or **networks**.
- In-house help desk professionals are often found in large companies, government departments, and universities, all of which generally have huge computer networks and many computer users. In-house help desk professionals respond mainly to telephone inquiries from staff members experiencing any kind of software or hardware problem, from a program that keeps **crashing** to a broken mouse.

A help desk professional often uses a computer to help resolve problems users may be having with software applications.

Sometimes, a help desk professional has to provide assistance in person.

The Computer Boom

In the early 1950s, few people in the world had heard of computers, much less used one. Today, few organizations anywhere do not have computers or do not rely on information technology in some form.

Sometimes, a help desk professional has to provide assistance in person.

Help desk professionals try to resolve most problems by phone, but some problems are too complex, especially when the caller has little technical knowledge or is not very confident using a computer. In these cases, in-house help desk professionals may go directly to a caller's work area, while help desk professionals who are not in-house will usually contact someone to assist at the work site.

Main responsibilities of a help desk professional

People contact help desk professionals in a number of ways. Most often, contact is made by phone, but questions can also be E-mailed, faxed, or discussed in person. It is important that help desk professionals are easy to contact so they can deal with problems quickly and efficiently.

Help desk professionals have to keep a record of all calls for assistance.

A help desk professional's main tasks are investigating computer problems and solving them.

The process of investigating a problem involves either

- asking the caller a series of questions to determine the nature of the problem or
- making a visit to the caller's location to figure out the problem.

Good Points and Bad Points

"I like the level of contact I have with people and enjoy being able to help them. The only drawback about my job is that some people are angry even before I receive their calls. The anger is usually due to their confusion and frustration. I try to make them happy again before the call ends."

After a help desk professional **diagnoses** a problem, he or she is responsible for finding a solution and guiding the caller through the solution. Help desk professionals guide callers in the following ways:

- giving step-by-step instructions over the phone
- sending detailed instructions by fax or E-mail
- going to the caller's location or workstation to provide personal assistance
- accessing the caller's computer system over the Internet, or through some other type of direct connection, to fix the problem

Help desk professionals record the problems from all of the calls they receive. By recording the subject of each call, they develop a knowledge base, which is a body of **data** about the kinds of problems that are occurring. A knowledge base can be used to **upgrade** and redesign products to get rid of common problems.

Besides helping people with their specific problems, many help desk professionals also provide general computer advice and training to computer users.

Main qualifications of a help desk professional

Technical knowledge

Help desk professionals have to know a great deal about the kinds of hardware and software callers are using. They must also know a lot about communicating and sharing information electronically.

Communication skills

Being able to ask questions and explain solutions clearly and simply is as important for a help desk professional as knowing the right questions to ask to diagnose a problem and the technical information to solve it. Often, callers will say they have tried something when they have not, so help desk professionals must be able to ask the same questions several different ways. Besides verbal skills, help desk professionals also need writing skills to record calls clearly and to make written instructions easy to understand.

Listening skills

Because they most often deal with callers who have less technical knowledge than they do, help desk professionals must be good listeners, and they have to be able to interpret what callers are saying.

The help desk areas in some companies are very large and employ a lot of people.

Quick thinking

In dealing with people who may be unhappy or angry, being able to identify problems and offer solutions quickly is very important.

Some help desk professionals set up computer systems as well as solve the problems people have using them.

fact file

Many help desk professionals get into the work after learning an organization's computer system while holding a different job within the organization, usually in the computer or IT area or department.

Some employers require a college degree or a certificate of computer training, while others are less concerned about these qualifications and provide their own help desk training.

Friendliness

Being friendly is essential, no matter how difficult the caller or the situation is. Even when callers are very upset or are not explaining their problems very clearly, help desk professionals must remain calm and tactful.

High standards

In some cases, the help desk is a person's main contact with an organization. Callers are more likely to have confidence in an organization when they feel they are being given high-quality, professional assistance.

A day in the life of a help desk professional

Kathryn Dean

Kathryn is a help desk professional for the offices of a large government agency.

8:00 a.m. I usually arrive at the office earlier than other members of the staff so I will be set up and ready before anyone starts having problems.

8:30 a.m. I put on my headset and answer incoming phone calls. As I listen to the calls, I type everything into my computer system to keep track of the kinds of things that seem to be going wrong.

11:00 a.m. In between phone calls, I check my E-mail for help requests. E-mail requests tend to be either a little less urgent than phone calls or a lot more complicated, which, from my point of view, makes answering E-mail requests more interesting. While I'm taking care of E-mail requests, I still have to answer the phone. I have to be able to keep all kinds of information in mind at the same time.

12:30 p.m. I grab a sandwich and eat while I study for a professional certification exam.

1:30 p.m. Having written rough drafts of answers to the more difficult problems I received by E-mail this morning, I now go over the drafts and send E-mail responses. While I'm working on the E-mails, I continue to answer incoming phone calls.

2:30 p.m. I go to see the people who need me to come to their workstations to take care of problems. They have either very complicated problems or a piece of computer equipment that needs fixing.

4:00 p.m. I stop at the computer lab to spend a little time learning a new software application. I also want to try to recreate a couple of puzzling problems from today's E-mails. I haven't been able to find an answer to these problems, but trying to create the problem myself often leads me to a solution.

5:30 p.m. People are leaving the building for the day so I can start thinking about going home.

Sometimes, even an on-site problem has a very simple solution.

Keeping the details of each problem straight can be difficult.

Software Developer

What is a software developer?

Software developers, who are often known as computer programmers, design and create computer programs, or software applications, which are the word processing, spread sheet, Web browser, desktop publishing, and other programs that run on computer systems.

Some software developers design programs that help computers communicate with each other as part of local or wide area **networks** (LANs or WANs). Computer networks can make communication within organizations much quicker and more effective.

There are two main types of software developers, those who create software to meet the needs of a particular business or organization and those who design software to be sold to the general public.

Software developers create computer programs for use in many different professions.

- An in-house software developer works for a particular organization, creating and maintaining software that is usually designed to support the organization's daily business operations, such as accounting programs for banks or research products for universities.
- An independent, or self-employed, software developer usually creates software for at least several different clients. Today, businesses in all areas of **commerce** and organizations of every kind and size rely on software, but they do not all employ in-house developers.

Bugs in the Software

Japanese researchers have shown that nearly 37 percent of the program errors, or bugs, discovered in new software **releases** could have been avoided if the developers had been given more time to design and test their products.

Independent software developers meet with representatives of companies to determine their business needs, then create, and sometimes maintain, software that meets those needs.

- Commercial software developers create software applications that are designed for general use by individuals, such as for home computers. These developers may be independent but, more often, are employed by software development companies. The kinds of software they create are usually referred to as end-user applications because they were not designed to meet the specific needs of any particular user.

Independent, or **freelance**, software developers may have to work at locations far from home.

Throughout the world, huge, multinational computer companies employ thousands of software developers. At one time, so many of these companies were located in one part of California that the area became known as Silicon Valley.

Main responsibilities of a software developer

Most software developers, whether working in-house or independently, are part of a team that includes other software developers as well as software **analysts**. The developers work alongside the analysts, looking at the problems new software is intended to solve and coming up with **cost-effective** solutions.

The process for developing most software programs includes the following tasks:

- getting to know a client's business, in detail, which means spending a lot of time with clients (Because independent software developers travel to their clients' business locations, they may have to work away from home for long periods of time.)
- presenting ideas to a client's **management**, usually explaining in great detail why one solution is better than any others

A software developer needs to be sure that clients know how proposed software will solve their business problems.

Good Points and Bad Points

"Working as part of a small team is great, and the feeling I get at the end of a project, when the software is up and running, is amazing. The only drawback I can think of is the long hours the team often has to work as we approach the deadline for completing a program."

- deciding on the exact requirements of a software application to best meet a client's needs
- writing the program **code** (Writing computer programs used to take a long time, but now, CASE (Computer Assisted Software Engineering) programs help the process, and they are often **automated**, which means that the CASE program translates the programmer's instructions into code.)
- testing the program to make sure it works the way it should on all of the client's computer systems

Developers often train clients how to use new software programs.

The final step in the software development process is writing instruction manuals and on-line help systems, which are usually done by technical writers as the software is being developed. Instructions for software applications must be written in a way that is easy to understand because many of the people who will be using the software may not be very familiar with computers or very comfortable using them.

Main qualifications of a software developer

Technical knowledge
Software developers must have a great deal of technical knowledge and be experts in the use of **hardware** and software. **Programming**, or the actual writing of the program code, is a big part of the job for many software developers. Writing program code requires expert knowledge of one or more **programming languages**.

Mathematical ability
Being good with numbers is important, especially for software developers involved in creating programs for research. Developers of research software often have to write programs that will carry out complicated mathematical operations.

Analytical ability
Examining a client's problems and figuring out exactly what the software will have to do for the client requires the ability to efficiently and effectively analyze even the most complex business problems and solutions.

Presentation skills
Convincing clients that a certain software solution is better than any other calls for a presentation that is better than any other. Presentation is important when giving instructions to users, too.

Working with computers means there is always something new to learn.

Writing the programming code is usually the most time-consuming part of a software developer's job.

fact file

More and more, any kind of job in computer technology is requiring a college degree in computer science or engineering, yet some software developers and other technologists have backgrounds as varied as physics, history, business administration, English, or even music. Interest and training are the keys, and because technology is constantly changing, lifelong learning and ongoing training are essential for technology-related careers.

Communication skills

Good communication is essential because software developers have to talk to their clients about what the clients want and how to best accomplish it. Many clients will even expect to receive detailed written **proposals**.

Teamwork

Most software developers work as part of a team that often includes people from many job areas. To achieve the best results, members of the group have to work well together.

A day in the life of a software developer

David Nugent

David is a senior software developer for a multinational software development company.

9:00 a.m. When I first arrive at work, I organize my work-station for the day. I usually have about half an hour to check my E-mails and get ready for my first meeting or appointment.

9:30 a.m. I attend a team meeting to discuss developments related to one of our projects. These meetings often involve looking at possible answers to new problems that have come up.

11:00 a.m. I meet with analysts and potential clients to discuss what the clients are looking for. Meetings like this one often go on longer than intended because clients don't always realize how involved what they're asking for can be.

1:00 p.m. I usually take half an hour for lunch, but I often have to eat on the run.

1:30 p.m. My next job is overseeing program construction. A team of programmers is writing the code. My job is to review the code that has been written and make sure programming is going well. I also spend part of the afternoon going over testing results with programmers to see how well new software is working during the testing phase.

3:30 p.m. I review the progress of the day and check for any E-mails I may have received since morning. I answer my E-mail messages and update my calendar of future appointments.

5:00 p.m. Time to pack up. Today has been fairly relaxed. When the team is approaching a deadline, I often have to work as late as 9:00 p.m.

Client meetings help software developers determine exactly what the new software will have to do.

Software developers have to stay in contact with many people.

Systems Analyst

Systems **analysts** look at business problems from both technical and **trade** points of view. Their jobs are to determine ways that computer systems can be developed and used to solve the problems.

Any organization that has a large computer **network**, whether it's a bank, a supermarket, a university, or a government department, needs the services of a systems analyst. Many large organizations employ one or more systems analysts in-house. Systems analysts who are self-employed work in their clients' offices for as long as it takes to solve the problems.

Whether in-house or independent, systems analysts work with people in many areas of a company.

The Importance of Computers

In 1990, business investments in computer systems, worldwide, amounted to about 20 percent of companies' total investment programs. By the year 2000, the amount of investment in computers had doubled to about 40 percent.

Systems analysts deal with many different types of businesses and business activities. One of the fastest growing types of business is **e-commerce**, which is the name given to buying and selling carried out over the Internet. An important part of an e-commerce systems analyst's job is making sure that clients are linked correctly to a company's internal networks so they are able to do business with each other.

The job of a systems analyst typically involves preparing a lot of reports and proposals.

On the job, systems analysts work with many different people, groups, and departments. Some of the main groups and departments are as follows:

- **management** teams (to discuss the goals of a project)
- information security teams (to make sure a new system is safe from **corruption**, tampering, or any other kind of interference)
- network operations (to check that a new system will work properly with an existing system or can replace the existing system)
- **software** developers (to guide them in creating new software or making the changes needed in a system's present software)

Main responsibilities of a systems analyst

Systems analysts create technological solutions to business problems, some of which may be quite complicated. Their jobs generally involve most, if not all, of the following tasks:

- drawing up detailed plans for an **upgraded** or new computer system
- creating a work plan that includes time for design, development, **installation**, and **integration**, with all phases completed by a specified deadline date
- calculating the cost of a project
- comparing a project's cost to its benefits to help determine whether the changes are worth making
- working with computer programmers, designers, and other skilled staff to make sure a system design will work well and will meet the client's business needs

Systems analysts often help train staff members in the use of a new computer system.

Good Points and Bad Points

"The part of my job I like most is working with the Internet security team. It is reassuring to know that, once I have created a new computer system, it is safe from attack by anyone outside the company. One drawback of the job is the amount of time some problems take to solve."

- developing a test plan to discover any problems in a system's operation that may not have been considered in the system's design
- managing the installation of a new system to make sure all of its features are working properly
- teaching a company's staff how to use a new or upgraded computer system, which can mean anything from personally conducting training sessions to simply making sure that instruction manuals are correctly written and printed

Installation projects can be large or small and are needed by organizations of all kinds.

Security has become an important part of IT work, and it often falls to systems analysts, working with security experts, to make sure that the information stored in a system is safe from **hackers** and other security risks.

Main qualifications of a systems analyst

Technical knowledge
To upgrade computer systems or create new ones, a systems analyst has to consider the computer equipment the system will run on. For this reason, the analyst has to know a lot about both **hardware** and software.

Logic
Analysts have to plan systems and investigate problems in a logical, step-by-step manner.

A project may go through several changes before its completion.

Business sense
A strong understanding of the businesses they work for is needed to determine the system requirements of companies and clients. Often, a self-employed systems analyst specializes in a particular type of business, such as banking or manufacturing.

Strict deadlines are often the driving force behind a systems analyst's work.

fact file

The most common way to become a systems analyst is to earn a bachelor's or master's degree in computer science or business, but companies often hire people with degrees that are not computer-related, then train them on the job.

The computer industry tends to be flexible about a systems analyst's educational background. Employers sometimes prefer systems-related work experience over any particular educational qualifications.

Communication skills

Whether they are speaking or writing, systems analysts must communicate clearly and be able to provide information in an understandable way, often to people who have little technical knowledge.

Patience

Analysts frequently meet with resistance or differences of opinion from clients about their plans for new computer systems or upgrades. These situations call for tact and patience, whether trying to persuade a client to accept a plan or working with the client to alter a plan.

Jasmeena Mistri

Jasmeena is twenty-seven years old and works as an in-house systems analyst.

9:00 a.m. I **boot up** my computer, read my E-mails, and use my calendar software to check my appointments for the day.

9:30 a.m. My first appointment is a meeting of the systems analysis team to review the plans for a financial institution's new sales network. At this meeting, we have to calculate the approximate costs of developing the system designs we have created.

11:30 a.m. I attend a meeting with financial analysts to discuss the possible costs of the project. We look at the cost of making each of two designs and compare it to the **profits** the company is likely to make. Financial analysts have a clearer idea than system analysts about the cost benefits of a project.

1:00 p.m. The meeting continues through a working lunch because the opinions of the financial analysts are needed for the next step of the project.

2:15 p.m. The systems analysis team meets to discuss possible changes to our designs. We have to try to make changes because the financial analysts feel that, without them, the project will be too expensive. The most common changes in plans for a system design are made to cut costs and keep the project within the client's budget.

4:45 p.m. I send E-mails describing suggested design changes to everyone involved in the project.

5:30 p.m. I review my schedule for tomorrow and gather up any materials I need to take home with me for preparation. Then, I turn off my computer and go home.

9:00 p.m. I put my children to bed and start preparing for tomorrow's meetings.

Working long hours is sometimes the only way a systems analyst can meet the deadline for having a new computer system up and running.

Systems analysts check that computers are operating well at every workstation.

Technical Sales Specialist

What is a technical sales specialist?

Technical sales specialists work for companies that produce either **hardware** or **software** products for computer systems. Their jobs involve selling their companies' products to information technology (IT) businesses, computer departments in other companies, or any organization that might use hardware or software.

The role of a technical sales specialist is similar, in most ways, to that of any salesperson, whether the products being sold are cars, vacuum cleaners, or vacation travel.

Sales goals are an important part of any technical sales specialist's work.

Targeting the Market

Individual computer users account for only about one-third of computer sales. The majority of computer purchases are made by businesses, government departments, universities, and schools. Most businesses have reached the point where every employee has a computer-equipped workstation.

Because technical sales specialists work with such advanced technology, however, it is very important that they fully understand their products and are able to explain the benefits of those products to clients.

Technical sales specialists sell products that can cost millions of dollars. An example might be software that a bank uses to **monitor cash flow** throughout its branches. The sales specialist representing the company that produces the software has to convince the bank's technical representatives that the product he or she is selling is better than the product currently in place and will save the bank money in the long run.

Clients typically want detailed cost estimates before deciding to buy expensive hardware or software products.

The cost of the products and the complicated jobs the products do mean that technical sales specialists have to work very hard to make a sale. Sometimes, the sales process for technical products goes on over a period of weeks or months, instead of hours or days.

Main responsibilities of a technical sales specialist

Technical sales specialists are often given an area or region in which they are responsible for selling their company's products. They may sell hardware, which is the computer equipment itself, or software, which are the programs that run on computers.

Most technical sales specialists spend a lot of time traveling.

The computer systems and products that technical sales specialists have to offer can be complicated. A computer system for an international chain of hotels, for example, is designed for booking hotel rooms and facilities around the world. To sell hardware and software products this complex, it is essential that a technical sales specialist knows his or her clients' businesses extremely well.

Good Points and Bad Points

"I am very much a people person, which is why I work in sales. In a sales job, I can meet and talk to lots of different people. Working in technical sales is particularly good for me because I can make use of my technical background and be sociable at the same time."

"Although I like the satisfaction of meeting my sales goals, I often have to work long hours and don't have much free time."

Technical sales specialists often have to spend nights away from home.

One of a technical sales specialist's most important tasks is building and maintaining a **client base** for his or her company within the assigned sales area. Sales specialists try to keep their customers loyal in the following ways:

- staying in regular contact with customers after a sale
- introducing new products to existing customers first
- providing preferential treatment and special offers

Sales specialists also have to keep expanding their client bases, which means bringing in new business for the company from within the sales region. Some of the ways technical sales specialists attract new clients include:

- contacting potential customers by phone, E-mail, or in person to introduce the company and its products
- arranging appointments to explain or demonstrate new hardware or software
- participating in trade shows, job fairs, or business expositions within the sales area or region

Main qualifications of a technical sales specialist

Technical knowledge

To sell sophisticated electronic products, technical sales specialists have to deal with technical staff so they must have an excellent understanding of information technology in general and their own products in particular. They must also be able to explain how their products perform better than products made by other companies.

Friendliness

Anyone in sales needs to have a warm, friendly approach and be able to get along well with all types of people.

Communication skills

Technical sales specialists routinely meet with clients to discuss the clients' needs and to present the benefits of the products they have to sell. To do their jobs well, sales specialists have to be able to speak clearly and in an interesting way and make complex information as simple as possible.

Business sense

Making a sale that benefits both buyer and seller takes good business sense, such as knowing the value of a product to a client and having a pretty good idea as to how much the client would be willing pay.

A sales specialist is always meeting with clients and has to know how to make them feel comfortable.

Discussing a client's needs is the first step toward making a sale.

Confidence

Success in sales work means knowing how to "close a deal," or, in other words, convincing the client to buy the product. To be convincing, a salesperson has to show confidence in the product as well as in his or her ability to sell it.

fact file

Many technical sales specialists have business or marketing degrees as well as technology-related training, but, sometimes, people move into sales after first working in hardware production or software design. Sales specialists with this kind of background usually know a lot about their companies' products.

Optimism

Sales work will always involve contact with people who decide not to buy the product. It is very important that sales specialists not let situations like these defeat them. Staying positive and cheerful is one of the best ways to maintain the confidence they need to be successful.

Alison O'Rourke

Alison is a technical sales specialist for a multinational software solutions company.

7:15 a.m.   I'm already on my way to a sales appointment that is about a hundred miles from my home.

9:30 a.m.   I arrive at my appointment about fifteen minutes early. I arrive early on purpose. There is no worse way to start a sales meeting than by being late. Being late for an appointment makes it look as if you're not very interested in the client.

9:45 a.m.   I give about a one-hour presentation to the client's **management** team to show what our software can do for the company. The presentation includes a twenty-minute demonstration of how the software works. After the presentation, I spend time with the company's technical staff, discussing the ways we can alter the software to run on the client's existing computer systems and add new **data fields** that better reflect the client's way of doing business.

12:00 p.m.   I have lunch with the client, and we talk about how much time it will take to **install** the software and train employees to use it.

Meetings held over lunch offer a more sociable than businesslike atmosphere.

To get and keep a client's attention, a sales specialist needs to be confident and offer an interesting and understandable presentation.

1:30 p.m.   I leave the client and drive back to my office.

2:45 p.m.   I arrive at the office, check my E-mails, and start returning missed phone calls.

3:30 p.m.   I meet with the software development team to tell them about the changes suggested during my morning meeting.

5:00 p.m.   I get in my car and drive home, taking paperwork from this morning's meeting with me. This evening, I will prepare a report on the development team's **assessment** of possible software changes. The report will include cost information that illustrates how much money the client can save. This report is called a **proposal**. I want to send the proposal out right away to show the client my interest.

Web Site Developer

What is a Web site developer?

Web site developers design and build pages on the World Wide Web for clients of all kinds. People all over the world can visit Web sites through the Internet.

The more visually attractive a Web site is, the more likely people will visit it.

In meetings with their clients, Web site developers determine the kinds of information and features clients need or want to display on the World Wide Web. A bank, for example, needs to show current interest rates. A movie theater will want to display show times and ticket prices. Web site developers not only have to determine the wording Web pages should contain but also design an attractive layout and build all of a site's features.

A Web site's appearance is usually decided by the type of business or organization it represents and the kinds of visitors that business or organization might attract. A government Web site, for example, needs to have a more formal look than a site for a music store.

Who Wants to Be a Billionaire?

The forty richest millionaires and billionaires in the United States made their money through the Internet, and every week in Australia, almost two hundred new millionaires are created through e-commerce and other Internet activities.

Some Web site developers work for Internet development companies or general design agencies. With an increasing number of businesses discovering the advantages of being on the Web, these kinds of companies usually have a wide variety of clients. Other Web site developers are self-employed and also may have any number of different clients. Some very large businesses have in-house Web site developers.

Besides creating new Web sites, developers also update existing sites. Customers frequently need additions and changes to their original sites. Their changes sometimes involve adding **e-commerce** features, such as on-line catalogs and the ability to make purchases on the site.

Frequent reports and on-line demonstrations help keep Web site clients up to date on the developer's progress.

Main responsibilities of a Web site developer

The first step in Web site development is meeting with clients to find out exactly what they expect from their sites. Some of the questions that need to be asked at these meetings include:

- Who is the site for?
- Why is the site needed?
- What ideas does the client have for the site?
- How much time does the developer have to create the site?
- How much money can the client spend?

Web site developers then design a site and create a draft of the design for the client to view.

Today, dozens of Web sites offer the convenience of booking air travel and other vacation plans on-line.

Good Points and Bad Points

"So many people have recently come to realize how much a Web site can help them that my job has become very interesting. These days, universities, travel agencies, hotels, and banks all have their own Web sites. Sometimes, however, it's hard to persuade clients to make the best use of their sites. Many of them don't understand everything that can be done on the Web. Also, many clients have a limited amount of money to spend on a site."

To write the **code** for a Web site, a developer has to decide on the best **programming language** to use. In choosing a programming language, most Web site developers consider the following:

- HTML, or hypertext markup language, is the classic Web-design **programming** code. It is well known and simple to use, but by today's standards, Web pages written only in HTML often appear plain.
- Javascript and other modern programming languages allow Web site developers to build more interesting sites and more complex page designs.

Clients must be given opportunities to view their Web sites while the sites are under construction. After seeing their sites on-line, they often change their minds about even their own ideas. Viewing a site as it is being built gives the client an early opportunity to tell the developer about anything that has to be changed.

When a site has been completed and approved by the client, the Web site developer has to **upload** the site to a **server**, which is a computer that allows **Web browsers access** to the site by way of the Internet.

Main qualifications of a Web site developer

Technical skills
High levels of computer knowledge and skills are essential for Web site developers. With programming languages becoming more and more powerful, Web site development has become increasingly complicated.

Creativity
Because Web site development companies need people who can design attractive Web pages, some of them prefer to hire people with more artistic ability than computer skills. In the future, artistic skills may be even more important because **software** is being developed to automatically write the code for Web sites, leaving more time for the creative process. Right now, however, most Web designers have to write the code themselves.

Developing Web sites is, most often, a team effort that may include writers and designers as well as computer programmers.

Writing ability
To write **proposals**, progress reports, and sometimes even Web page content, Web site developers have to be good writers.

Web sites for children are usually designed to be both entertaining and educational.

Presentation skills

Making presentations is a big part of a Web site developer's job. In meetings to sell ideas to clients or present drafts of Web pages for review, they need to speak confidently and present information in a clear and interesting way.

Teamwork

Many different talents go into building Web sites so Web developers have to be able to work well as part of a team.

fact file

Although growing quickly, Web site development is a fairly new profession. Many developers work **freelance**, and companies that hire developers are often flexible about qualifications. They may be looking for college graduates, but, more important, they want creative people who are able to learn new skills and keep up with new technology.

A day in the life of a Web site developer

Paula James

Paula is the lead Web site developer for a small Web development company.

7:30 a.m. I like to get to work early because Web design projects seem to have problems that spring up from all directions. Before I start working on a project, I read my E-mails and plan my day.

9:00 a.m. I meet with the design team. There are four of us, and today, we need to bring each other up to date on the progress we've made in completing our individual tasks for a Web site plan that I have to present to a client this afternoon.

Our programmer has some ideas on **interactive** features for the site. We spend time discussing how to present these ideas to the client.

10:30 a.m. In the next two hours, I have to make all of the last-minute changes the team agreed to at this morning's meeting. There are a lot of them.

12:30 p.m. Lunchtime.

1:15 p.m. I present the Web site plan to our client. As the team's leader, I am usually the one responsible for presenting draft sites to customers.

A draft site is a simplified example of a client's final site. It gives the customer an idea of how the finished site will look and operate. Giving draft site presentations is one of the most important parts of my job. I have to make sure the customer understands exactly what is happening on the site and feels that we are doing a good job.

3:00 p.m. The team meets again to discuss the outcome of my presentation to the client. These post-presentation meetings have been known to last as long as five hours when there are a lot of changes needed to meet the client's wishes.

The client liked the ideas for interactivity but asked for some **enhancements**. Most of the meeting was spent discussing these issues.

5:30 p.m. The post-presentation meeting was pretty short, but I've still had a long day, so I quickly check my E-mails, tidy up my office, and go home.

Besides being happy with the page designs, clients need to understand how their Web sites will operate.

Sharing ideas and talents is one of the rewards of designing Web sites.

Glossary

access – the ability to enter, make use of, or communicate with

analysts – people who examine problems or situations by breaking them down to find solutions or come to conclusions

assessment – evaluation or appraisal

automated – set up to do something mechanically, without human assistance

boot up – to start a computer, preparing the installed applications for use

brainstorming – generating ideas rapidly, without stopping to evaluate them

cash flow – the movement of money coming in to (income, or earnings) and going out of (expenses) a business

client base – the people or organizations that do business with a company on a regular basis

code – the particular arrangements of numbers or characters of a programming language that create computer programs and software applications

commerce – business, especially business involved in large-scale buying and selling

compatibility – the ability of separate pieces of equipment to work together

corruption – contamination or damage to what was original and intended

cost-effective – having economic benefits that justify money spent

crashing – suddenly stopping or ceasing to work while being used

data – factual information, such as measurements and statistics, especially as stored in a computer

database – a collection of particular and similar data stored, electronically, in a computer system

diagnose – to identify the cause by looking at the signs of a problem

e-commerce – commerce, or business, conducted over the Internet

enhancements – added features that improve quality, value, or appearance

fields – the areas in databases where similar types of information are stored

floppy disk drive – the part of a computer into which disks are inserted to allow the transfer of information

freelance – self-employed and free to work for more than one client

hackers – people who break into the computer systems and networks of others

hardware – the physical equipment and parts of a computer system

innovations – new ideas or methods

installation – the process of putting a computer system in place and connecting it for use or adding hardware or software to a system that already exists

integration – the process of getting computer systems to communicate or work with each other

interactive – capable of two-way communication

management – the group of people in an organization who are responsible for supervising work and making decisions

modem – a device that connects one computer to another by means of a telephone or cable line

monitor – (n) a video display device; (v) to check on or keep track of

network – a group of computers that are linked so they can communicate with each other directly and share information

PC-based – designed for a personal computer platform, such as Windows

platforms – computer operating systems, such as Windows, Macintosh, or UNIX

premises – areas of land and the buildings on them

profits – money made by selling a product or service for more than the cost of producing it

programming – the act of creating, or writing, the code that creates a computer program or software application

programming language – a system of defined numbers and characters used to give instructions to a machine

proposals – formal reports describing plans of action, which are presented to clients or decision-makers to be accepted or rejected

releases – newly developed software or new versions of existing software

server – a main computer within a network that organizes contact between the computers that form the network

shifts – scheduled periods of time when certain groups of workers are on duty

software – computer programs that give a computer the instructions to perform specific functions

trade – the dealings of business or industry, particularly buying and selling

upgrades – new pieces of hardware or new versions of software added to improve or update computer systems

upload – the process of transferring information to the Internet or some other place in a network

user interface – the means by which a person communicates with a computer, such as a keyboard, a mouse, or on-screen menu commands

virus – unauthorized computer code that spreads from one computer to another, destroying information or damaging operating systems

Web browsers – software designed for accessing the Internet and its Web sites

Further Information

This book does not cover all of the jobs that involve working with computers. Many jobs are not mentioned, including database developer, security specialist, and systems technician. This book does, however, give you an idea of what working with computers is like.

Computers have changed the world in a very short period of time. Today, there is a growing number of jobs in the information technology (IT) or computer industry, both for companies that specialize in IT and for IT departments in businesses, universities, hospitals, hotels, and many other kinds of organizations. This book tries to show the variety and range of computer-related jobs as well as make it clear that women are as successful working with computers as men.

The way to decide if working with computers is right for you is to find out what the work involves. Read as much as you can about computer-related careers and talk to people, especially people you know, who work with computers.

When you are in middle school or high school, a teacher or career counselor might be able to help you arrange some work experience in a certain career. For careers working with computers, that experience could mean spending some time at a help desk or in a computer lab, watching what goes on and how people who work there spend their time.

Books

Careers on the Web
Linda Bullock
(Raintree/Steck-
　Vaughn, 2003)

Computer Engineer
Melissa Maupin
(Capstone Press, 2001)

Computer Programmer
Peggy J. Parks
(Kidhaven, 2003)

Software Designer
Alice McGinty
(Rosen, 2003)

Web Sites

*The Education of a
　Computer Game
　Designer*
www.erasmatazz.com/
　library/Game%20
　Design/The_
　Education_of_a_
　Game_Designer.html

*GetTech Careers:
　Information Technology*
www.GetTech.org/
　careers.asp?category=3

*IEEE Computer Society:
　Careers in Computing*
www.computer.org/
　education/careers.htm

Useful Addresses

Games Designer

DigiPen Institute of Technology
5001 – 150th Avenue, NE
Redmond, WA 98052
Tel: (425) 558-0200
www.digipen.edu

International Game Developers Association
600 Harrison Street
San Francisco, CA 94107
Tel: (415) 947-6235
www.igda.org/students

Hardware Engineer

Association for Computing Machinery
1515 Broadway
New York, NY 10036
Tel: (800) 342-6626 or (212) 626-0500
www.acm.org

Help Desk Professional

Help Desk Institute
6385 Corporate Drive, Suite 301
Colorado Springs, CO 80919
Tel: (800) 248-5667 or (719) 268-0174
E-mail: support@thinkhdi.com
www.thinkhdi.com

Software Developer

IEEE Computer Society
10662 Los Vaqueros Circle
P. O. Box 3014
Los Alamitos, CA 90720-1314
Tel: (714) 821-8380
www.computer.org

Software & Information
  Industry Association
1090 Vermont Avenue, NW, sixth floor
Washington, DC 20005
Tel: (202) 289-7442
www.spa.org

Systems Analyst

Network and Systems Professionals
  Association, Inc.
7044 South 13th Street
Oak Creek, WI 53154
Tel: (414) 768-8000
www.naspa.com

Technical Sales Specialist

Business Technology Association
12411 Wornall Road, Suite 200
Kansas City, MO 64145
Tel: (816) 941-3100
bta.org

Web Site Developer

International Webmasters Association
119 E. Union Street, Suite F
Pasadena, CA 91103
Tel: (626) 449-8308
www.iwanet.org

World Organization of Webmasters
9580 Oak Avenue Parkway,
  Suite 7-177
Folsom, CA 95630
Tel: (916) 608-1597
E-mail: info@joinwow.org
www.joinwow.org

Index